# Quantum Physics

## FOR

## DUMMIES®

by Steven Holzner

WILEY

Wiley Publishing, Inc.

**Quantum Physics For Dummies®**

Published by
Wiley Publishing, Inc.
111 River St.
Hoboken, NJ 07030-5774
www.wiley.com

WILEY

# About the Author

**Steven Holzner** is an award-winning author of technical and science books (like *Physics For Dummies* and *Differential Equations For Dummies*). He graduated from MIT and did his PhD in physics at Cornell University, where he was on the teaching faculty for 10 years. He's also been on the faculty of MIT. Steve also teaches corporate groups around the country.

# Author's Acknowledgments

I'd particularly like to thank the people at Wiley: Tracy Boggier, Tim Gallan, and Danielle Voirol.

# Dedication

To Nancy, of course!

## Publisher's Acknowledgments

Weíre proud of this book; please send us your comments through our Dummies online registration form located at www.dummies.com/register/.

Some of the people who helped bring this book to market include the following:

*Acquisitions, Editorial, and Media Development*

**Senior Project Editor:** Tim Gallan

**Acquisitions Editor:** Tracy Boggier

**Senior Copy Editor:** Danielle Voirol

**Assistant Editor:** Erin Calligan Mooney

**Technical Editor:** Dan Funch Wohns

**Editorial Manager:** Michelle Hacker

**Editorial Assistants:** Joe Niesen, Jennette ElNaggar, David Lutton

**Cartoons:** Rich Tennant (www.the5thwave.com)

*Composition Services*

**Project Coordinator:** Erin Smith

**Layout and Graphics:** Carl Byers, Carrie A. Cesavice, Shawn Frazier, Nikki Gately, Melissa Smith, Christine Williams

**Proofreaders:** Joni Heredia, Shannon Ramsey

**Indexer:** Broccoli Information Management

---

**Publishing and Editorial for Consumer Dummies**

> **Diane Graves Steele,** Vice President and Publisher, Consumer Dummies

> **Joyce Pepple,** Acquisitions Director, Consumer Dummies

> **Kristin Ferguson-Wagstaffe,** Product Development Director, Consumer Dummies

> **Ensley Eikenburg,** Associate Publisher, Travel

> **Kelly Regan,** Editorial Director, Travel

**Publishing for Technology Dummies**

> **Andy Cummings,** Vice President and Publisher, Dummies Technology/General User

**Composition Services**

> **Gerry Fahey,** Vice President of Production Services

> **Debbie Stailey,** Director of Composition Services

# Contents at a Glance

# Table of Contents

## Part II: Bound and Undetermined: Handling Particles in Bound States............................... 55

### Chapter 3: Getting Stuck in Energy Wells ......................57

# Introduction

●●●●●●●●●●●●●●●●●●●●●●●●●●●●●●●●●●●●●●●●●●●●●●●●●●●●●●●●●●●●●●●

*P*hysics as a general discipline has no limits, from the very huge (galaxy-wide) to the very small (atoms and smaller). This book is about the very small side of things — that's the specialty of quantum physics. When you *quantize* something, you can't go any smaller; you're dealing with discrete units.

Classical physics is terrific at explaining things like heating cups of coffee or accelerating down ramps or cars colliding, as well as a million other things, but it has problems when things get very small. Quantum physics usually deals with the micro world, such as what happens when you look at individual electrons zipping around. For example, electrons can exhibit both particle and wave-like properties, much to the consternation of experimenters — and it took quantum physics to figure out the full picture.

Quantum physics also introduced the uncertainty principle, which says you can't know a particle's exact position and momentum at the same time. And the field explains the way that the energy levels of the electrons bound in atoms work. Figuring out those ideas all took quantum physics, as physicists probed ever deeper for a way to model reality. Those topics are all coming up in this book.

## About This Book

Because uncertainty and probability are so important in quantum physics, you can't fully appreciate the subject without getting into calculus. This book presents the need-to-know concepts, but you don't see much in the way of thought experiments that deal with cats or parallel universes. I focus on the math and how it describes the quantum world.

I've taught physics to many thousands of students at the university level, and from that experience, I know most of them share one common trait: Confusion as to what they did to deserve such torture.

*Quantum Physics For Dummies* largely maps to a college course, but this book is different from standard texts. Instead of writing it from the physicist's or professor's point of view, I've tried to write it from the reader's point of view. In other words, I've designed this book to be crammed full of the good stuff — and only the good stuff. Not only that, but you can discover ways of looking at things that professors and teachers use to make figuring out problems simple.

Although I encourage you to read this book from start to finish, you can also leaf through this book as you like, reading the topics that you find interesting. Like other *For Dummies* books, this one lets you skip around as you like as much as possible. You don't have to read the chapters in order if you don't want to. This is your book, and quantum physics is your oyster.

# Conventions Used in This Book

Some books have a dozen dizzying conventions that you need to know before you can even start. Not this one. Here's all you need to know:

- ✔ I put new terms in italics, like *this,* the first time they're discussed; I follow them with a definition.
- ✔ Vectors — those items that have both a magnitude and a direction — are given in bold, like this: **B.**
- ✔ Web addresses appear in `monofont`.

# Foolish Assumptions

I don't assume that you have any knowledge of quantum physics when you start to read this book. However, I do make the following assumptions:

- ✔ You're taking a college course in quantum physics, or you're interested in how math describes motion and energy on the atomic and subatomic scale.
- ✔ You have some math prowess. In particular, you know some calculus. You don't need to be a math pro, but you should know how to perform integration and deal with differential equations. Ideally, you also have some experience with Hilbert space.
- ✔ You have some physics background as well. You've had a year's worth of college-level physics (or understand all that's in *Physics For Dummies*) before you tackle this one.

# How This Book Is Organized

Quantum physics — the study of very small objects — is actually a very big topic. To handle it, quantum physicists break the world down into different parts. Here are the various parts that are coming up in this book.

## Part I: Small World, Huh? Essential Quantum Physics

Part I is where you start your quantum physics journey, and you get a good overview of the topic here. I survey quantum physics and tell you what it's good for and what kinds of problems it can solve. You also get a good foundation in the math that you need for the rest of the book, such as state vectors and quantum matrix manipulations. Knowing this stuff prepares you to handle the other parts.

## Part II: Bound and Undetermined: Handling Particles in Bound States

Particles can be trapped inside potentials; for instance, electrons can be bound in an atom. Quantum physics excels at predicting the energy levels of particles bound in various potentials, and that's what Part II covers. You see how to handle particles bound in square wells and in harmonic oscillators.

## Part III: Turning to Angular Momentum and Spin

Quantum physics lets you work with the micro world in terms of the angular momentum of particles, as well as the spin of electrons. Many famous experiments — such as the Stern-Gerlach experiment, in which beams of particles split in magnetic fields — are understandable only in terms of quantum physics, and you get all the details here.

## Part IV: Multiple Dimensions: Going 3D with Quantum Physics

In the first three parts, all the quantum physics problems are one-dimensional to make life a little easier while you're understanding how to solve those problems. In Part IV, you branch out to working with three-dimensional problems in both rectangular and spherical coordinate systems. Taking things from 1D to 3D gives you a better picture of what happens in the real world.

## Part V: Group Dynamics: Introducing Multiple Particles

In this part, you work with multiple-particle systems, such as atoms and gases. You see how to handle many electrons in atoms, particles interacting with other particles, and particles that scatter off other particles.

Dealing with multiple particles is all another step in modeling reality — after all, systems with only a single particle don't take you very far in the real world, which is built of mega, mega systems of particles. In Part V, you see how quantum physics can handle the situation.

## Part VI: The Part of Tens

You see the Part of the Tens in all *For Dummies* books. This part is made up of fast-paced lists of ten items each. You get to see some of the ten best online tutorials on quantum physics and a discussion of quantum physics' ten greatest triumphs.

# Icons Used in This Book

You find a handful of icons in this book, and here's what they mean:

This icon flags particularly good advice, especially when you're solving problems.

This icon marks something to remember, such as a law of physics or a particularly juicy equation.

This icon means that what follows is technical, insider stuff. You don't have to read it if you don't want to, but if you want to become a quantum physics pro (and who doesn't?), take a look.

This icon helps you avoid mathematical or conceptual slip-ups.

# Where to Go from Here

All right, you're all set and ready to go. You can jump in anywhere you like. For instance, if you're sure electron spin is going to be a big topic of conversation at a party this weekend, check out Chapter 6. And if your upcoming vacation to Geneva, Switzerland, includes a side trip to your new favorite particle accelerator — the Large Hadron Collider — you can flip to Chapter 12 and read up on scattering theory. But if you want to get the full story from the beginning, jump into Chapter 1 first — that's where the action starts.

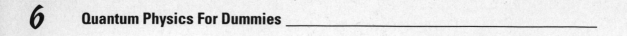

# Part I
# Small World, Huh? Essential Quantum Physics

The 5th Wave          By Rich Tennant

"It's just like the regular stew only it's got some bits of matter in it we can't identify."

# In this part . . .

This part is designed to give you an introduction to the ways of quantum physics. You see the issues that gave rise to quantum physics and the kinds of solutions it provides. I also introduce you to the kind of math that quantum physics requires, including the notion of state vectors.

# Chapter 1

# Discoveries and Essential Quantum Physics

In This Chapter

▶ Putting forth theories of quantization and discrete units

▶ Experimenting with waves acting as particles

▶ Experimenting with particles acting as waves

▶ Embracing uncertainty and probability

According to classical physics, particles are particles and waves are waves, and never the twain shall mix. That is, particles have an energy E and a momentum vector $p$, and that's the end of it. And waves, such as light waves, have an amplitude A and a wave vector $k$ (where the magnitude of $k = \frac{2\pi}{\lambda}$, where $\lambda$ is the wavelength) that points in the direction the wave is traveling. And that's the end of that, too, according to classical physics.

But the reality is different — particles turn out to exhibit wave-like properties, and waves exhibit particle-like properties as well. The idea that waves (like light) can act as particles (like electrons) and vice versa was the major revelation that ushered in quantum physics as such an important part of the world of physics. This chapter takes a look at the challenges facing classical physics around the turn of the 20th century — and how quantum physics gradually came to the rescue. Up to that point, the classical way of looking at physics was thought to explain just about everything. But as those pesky experimental physicists have a way of doing, they came up with a bunch of experiments that the theoretical physicists couldn't explain.

That made the theoretical physicists mad, and they got on the job. The problem here was the microscopic world — the world that's too tiny to see. On the larger scale, classical physics could still explain most of what was going on — but when it came to effects that depended on the micro-world, classical physics began to break down. Taking a look at how classical physics collapsed gives you an introduction to quantum physics that shows why people needed it.

# Being Discrete: The Trouble with Black-Body Radiation

One of the major ideas of quantum physics is, well, *quantization* — measuring quantities in discrete, not continuous, units. The idea of quantized energies arose with one of the earliest challenges to classical physics: the problem of black-body radiation.

When you heat an object, it begins to glow. Even before the glow is visible, it's radiating in the infrared spectrum. The reason it glows is that as you heat it, the electrons on the surface of the material are agitated thermally, and electrons being accelerated and decelerated radiate light.

Physics in the late 19th and early 20th centuries was concerned with the spectrum of light being emitted by black bodies. A *black body* is a piece of material that radiates corresponding to its temperature — but it also absorbs and reflects light from its surroundings. To make matters easier, physics postulated a black body that reflected nothing and absorbed all the light falling on it (hence the term *black body,* because the object would appear perfectly black as it absorbed all light falling on it). When you heat a black body, it would radiate, emitting light.

Well, it was hard to come up with a physical black body — after all, what material absorbs light 100 percent and doesn't reflect anything? But the physicists were clever about this, and they came up with the hollow cavity you see in Figure 1-1, with a hole in it.

When you shine light on the hole, all that light would go inside, where it would be reflected again and again — until it got absorbed (a negligible amount of light would escape through the hole). And when you heated the hollow cavity, the hole would begin to glow. So there you have it — a pretty good approximation of a black body.

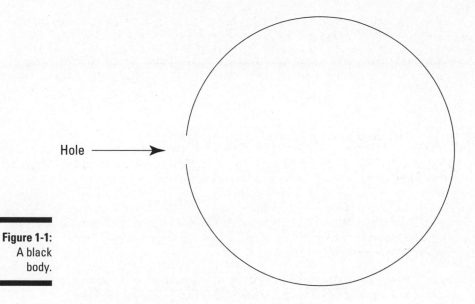

Hole →

**Figure 1-1:**
A black
body.

You can see the spectrum of a black body (and attempts to model that spectrum) in Figure 1-2, for two different temperatures, $T_1$ and $T_2$. The problem was that nobody was able to come up with a theoretical explanation for the spectrum of light generated by the black body. Everything classical physics could come up with went wrong.

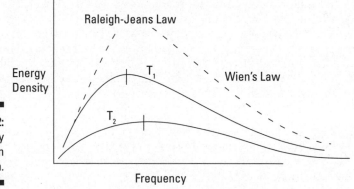

Raleigh-Jeans Law

Energy
Density

$T_1$

Wien's Law

$T_2$

**Figure 1-2:**
Black-body
radiation
spectrum.

Frequency

## First attempt: Wien's Formula

The first one to try to explain the spectrum of a black body was Willhelm Wien, in 1889. Using classical thermodynamics, he came up with this formula:

$$u(v, T) = Av^3 e^{-\beta v/T}$$

where A and β are constants you determine from your physical setup, $v$ is the frequency of the light, and T is the temperature of the black body. (The spectrum is given by $u[v, T]$, which is the energy density of the emitted light as a function of frequency and temperature.)

This equation, Wien's formula, worked fine for high frequencies, as you can see in Figure 1-2; however, it failed for low frequencies.

## Second attempt: Raleigh-Jeans Law

Next up in the attempt to explain the black-body spectrum was the Raleigh-Jeans Law, introduced around 1900. This law predicted that the spectrum of a black body was

$$u(v, T) = \frac{8\pi v^2}{c^3} kT$$

where $k$ is Boltmann's constant (approximately $1.3807 \times 10^{-23}$ J·K⁻¹). However, the Raleigh-Jeans Law had the opposite problem of Wien's law: Although it worked well at low frequencies (see Figure 1-2), it didn't match the higher-frequency data at all — in fact, it diverged at higher frequencies. This was called the *ultraviolet catastrophe* because the best predictions available diverged at high frequencies (corresponding to ultraviolet light). It was time for quantum physics to take over.

## An intuitive (quantum) leap: Max Planck's spectrum

The black-body problem was a tough one to solve, and with it came the first beginnings of quantum physics. Max Planck came up with a radical suggestion — what if the amount of energy that a light wave can exchange with matter wasn't continuous, as postulated by classical physics, but *discrete?* In other words, Planck postulated that the energy of the light emitted from the walls of the black-body cavity came only in integer multiples like this, where $h$ is a universal constant:

$$E = nh\upsilon, \text{ where } n = 0, 1, 2, \ldots$$

With this theory, crazy as it sounded in the early 1900s, Planck converted the continuous integrals used by Raleigh-Jeans to discrete sums over an infinite number of terms. Making that simple change gave Planck the following equation for the spectrum of black-body radiation:

$$u(\upsilon, \mathrm{T}) = \frac{8\pi\upsilon^2}{c^3} \frac{h\upsilon}{e^{h\upsilon/k\mathrm{T}} - 1}$$

This equation got it right — it exactly describes the black-body spectrum, both at low and high (and medium, for that matter) frequencies.

This idea was quite new. What Planck was saying was that the energy of the radiating oscillators in the black body couldn't take on just any level of energy, as classical physics allows; it could take on only specific, *quantized* energies. In fact, Planck hypothesized that that was true for *any* oscillator — that its energy was an integral multiple of $h\upsilon$.

And so Planck's equation came to be known as *Planck's quantization rule,* and $h$ became *Planck's constant: $h = 6.626 \times 10^{-34}$* Joule-seconds. Saying that the energy of all oscillators was quantized was the birth of quantum physics.

One has to wonder how Planck came up with his theory, because it's not an obvious hypothesis. Oscillators can oscillate only at discrete energies? Where did that come from? In any case, the revolution was on — and there was no stopping it.

# The First Pieces: Seeing Light as Particles

Light as particles? Isn't light made up of waves? Light, it turns out, exhibits properties of both waves and particles. This section shows you some of the evidence.

## Solving the photoelectric effect

The photoelectric effect was one of many experimental results that made up a crisis for classical physics around the turn of the 20th century. It was also one of Einstein's first successes, and it provides proof of the quantization of light. Here's what happened.

When you shine light onto metal, as Figure 1-3 shows, you get emitted electrons. The electrons absorb the light you shine, and if they get enough energy, they're able to break free of the metal's surface. According to classical physics, light is just a wave, and it can exchange any amount of energy with the metal. When you beam light on a piece of metal, the electrons in the metal should absorb the light and slowly get up enough energy to be emitted from the metal. The idea was that if you were to shine more light onto the metal, the electrons should be emitted with a higher kinetic energy. And very weak light shouldn't be able to emit electrons at all, except in a matter of hours.

But that's not what happened — electrons were emitted as soon as someone shone light on the metal. In fact, no matter how weak the intensity of the incident light (and researchers tried experiments with such weak light that it should have taken hours to get any electrons emitted), electrons *were* emitted. Immediately.

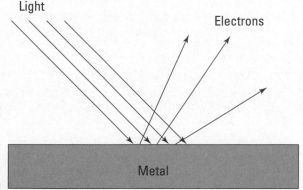

**Figure 1-3:**
The photo-
electric
effect.

Experiments with the photoelectric effect showed that the kinetic energy, K, of the emitted electrons depended only on the frequency — not the intensity — of the incident light, as you can see in Figure 1-4.

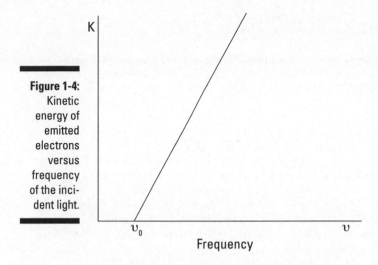

In Figure 1-4, $\upsilon_0$ is called the *threshold frequency*, and if you shine light with a frequency below this threshold on the metal, no electrons are emitted. The emitted electrons come from the pool of free electrons in the metal (all metals have a pool of free electrons), and you need to supply these electrons with an energy equivalent to the metal's work function, W, to emit the electron from the metal's surface.

The results were hard to explain classically, so enter Einstein. This was the beginning of his heyday, around 1905. Encouraged by Planck's success (see the preceding section), Einstein postulated that not only were oscillators quantized but so was light — into discrete units called *photons*. Light, he suggested, acted like particles as well as waves.

So in this scheme, when light hits a metal surface, photons hit the free electrons, and an electron completely absorbs each photon. When the energy, $h\upsilon$, of the photon is greater than the work function of the metal, the electron is emitted. That is,

$$h\upsilon = W + K$$

where W is the metal's work function and K is the kinetic energy of the emitted electron. Solving for K gives you the following:

$$K = h\upsilon - W$$

You can also write this in terms of the threshold frequency this way:

$$K = h(\upsilon - \upsilon_0)$$

So apparently, light isn't just a wave; you can also view it as a particle, the photon. In other words, light is quantized.

That was also quite an unexpected piece of work by Einstein, although it was based on the earlier work of Planck. Light *quantized*? Light coming in discrete energy packets? What next?

## Scattering light off electrons: The Compton effect

To a world that still had trouble comprehending light as particles (see the preceding section), Arthur Compton supplied the final blow with the Compton effect. His experiment involved scattering photons off electrons, as Figure 1-5 shows.

**Figure 1-5:** Light incident on an electron at rest.

Photon $\lambda$

Electron at rest

Incident light comes in with a wavelength of $\lambda$ and hits the electron at rest. After that happens, the light is scattered, as you see in Figure 1-6.

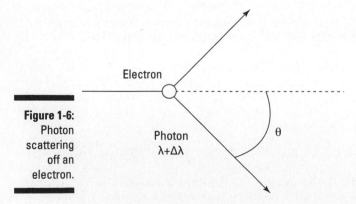

Electron

**Figure 1-6:** Photon scattering off an electron.

Photon $\lambda + \Delta\lambda$

$\theta$

Classically, here's what should've happened: The electron should've absorbed the incident light, oscillated, and emitted it — with the same wavelength but with an intensity depending on the intensity of the incident light. But that's not what happened — in fact, the wavelength of the light is actually changed by $\Delta\lambda$, called the *wavelength shift*. The scattered light has a wavelength of $\lambda + \Delta\lambda$ — in other words, its wavelength has increased, which means the light has lost energy. And $\Delta\lambda$ depends on the scattering angle, $\theta$, not on the intensity of the incident light.

Arthur Compton could explain the results of his experiment only by making the assumption that he was actually dealing with two particles — a photon and an electron. That is, he treated light as a discrete particle, not a wave. And he made the assumption that the photon and the electron collided elastically — that is, that both total energy and momentum were conserved.

Making the assumption that both the light and the electron were particles, Compton then derived this formula for the wavelength shift (it's an easy calculation if you assume that the light is represented by a photon with energy $E = h\upsilon$ and that its momentum is $p = {}^E/_c$):

$$\Delta\lambda = \frac{h}{m_e c}\left(1 - \cos\theta\right)$$

where $h$ is Planck's constant, $m_e$ is the mass of an electron, $c$ is the speed of light, and $\theta$ is the scattering angle of the light.

You also see this equation in the equivalent form:

$$\Delta\lambda = 4\pi\lambda_c \sin^2\left(\theta/_2\right)$$

where $\lambda_c$ is the Compton wavelength of an electron, $\lambda_c = \hbar/m_e c$, where $\hbar = h/2\pi$. And experiment confirms this relation — both equations.

Note that to derive the wavelength shift, Compton had to make the assumption that here, light was acting as a particle, not as a wave. That is, the particle nature of light was the aspect of the light that was predominant.

# Proof positron? Dirac and pair production

In 1928, the physicist Paul Dirac posited the existence of a positively charged anti-electron, the *positron*. He did this by taking the newly evolving field of quantum physics to new territory by combining relativity with quantum mechanics to create relativistic quantum mechanics — and that was the theory that predicted, through a plus/minus-sign interchange — the existence of the positron.

It was a bold prediction — an *anti-particle* of the electron? But just four years later, physicists actually saw the positron. Today's high-powered elementary particle physics has all kinds of synchrotrons and other particle accelerators to create all the elementary particles they need, but in the early 20th century, this wasn't always so.

In those days, physicists relied on cosmic rays — those particles and high-powered photons (called gamma rays) that strike the Earth from outer space — as their source of particles. They used *cloud-chambers*, which were filled with vapor from dry ice, to see the trails such particles left. They put their chambers into magnetic fields to be able to measure the momentum of the particles as they curved in those fields.

In 1932, a physicist noticed a surprising event. A pair of particles, oppositely charged (which could be determined from the way they curved in the magnetic field) appeared from apparently nowhere. No particle trail led to the origin of the two particles that appeared. That was *pair-production* — the conversion of a high-powered photon into an electron and positron, which can happen when the photon passes near a heavy atomic nucleus.

So experimentally, physicists had now seen a photon turning into a pair of particles. Wow. As if everyone needed more evidence of the particle nature of light. Later on, researchers also saw *pair annihilation:* the conversion of an electron and positron into pure light.

Pair production and annihilation turned out to be governed by Einstein's newly introduced theory of relativity — in particular, his most famous formula, $E = mc^2$, which gives the pure energy equivalent of mass. At this point, there was an abundance of evidence of the particle-like aspects of light.

# A Dual Identity: Looking at Particles as Waves

In 1923, the physicist Louis de Broglie suggested that not only did waves exhibit particle-like aspects but the reverse was also true — all material particles should display wave-like properties.

How does this work? For a photon, momentum $p = {}^{h\upsilon}/_c = {}^h/_\lambda$, where $\upsilon$ is the photon's frequency and $\lambda$ is its wavelength. And the wave vector, **k,** is equal to $\boldsymbol{k} = \boldsymbol{p}/\hbar$, where $\hbar = h/2\pi$. De Broglie said that the same relation should hold for all material particles. That is,

$$\lambda = \frac{h}{p}$$

$$k = \frac{p}{\hbar}$$

De Broglie presented these apparently surprising suggestions in his Ph.D. thesis. Researchers put these suggestions to the test by sending a beam through a dual-slit apparatus to see whether the electron beam would act like it was made up of particles or waves. In Figure 1-7, you can see the setup and the results.

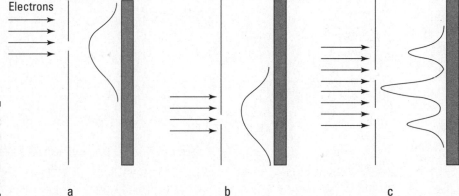

**Figure 1-7:**
An electron beam going through two slits.

a                    b                    c

In Figure 1-7a, you can see a beam of electrons passing through a single slit and the resulting pattern on a screen. In Figure 1-7b, the electrons are passing through a second slit. Classically, you'd expect the intensities of Figure 1-7a and 1-7b simply to add when both slits are open:

$$I = I_1 + I_2$$

But that's not what happened. What actually appeared was an interference pattern when both slits were open (Figure 1-7c), not just a sum of the two slits' electron intensities.

The result was a validation of de Broglie's invention of matter waves. Experiment bore out the relation that $\lambda = {}^h/_p$, and de Broglie was a success.

The idea of matter waves is a big part of what's coming up in the rest of the book. In particular, the existence of matter waves says that you add the waves' amplitude, $\psi_1(r, t)$ and $\psi_2(r, t)$, not their intensities, to sum them:

$$\psi(r, t) = \psi_1(r, t) + \psi_2(r, t)$$

You square the amplitude to get the intensity, and the phase difference between $\psi_1(r, t)$ and $\psi_2(r, t)$ is what actually creates the interference pattern that's observed.

# You Can't Know Everything (But You Can Figure the Odds)

So particles apparently exhibit wave-like properties, and waves exhibit particle-like properties. But if you have an electron, which is it — a wave or a particle? The truth is that physically, an electron is just an electron, and you can't actually say whether it's a wave or a particle. The act of *measurement* is what brings out the wave or particle properties. You see more about this idea throughout the book.

Quantum mechanics lives with an uncertain picture quite happily. That view offended many eminent physicists of the time — notably Albert Einstein, who said, famously, "God does not play dice." In this section, I discuss the idea of uncertainty and how quantum physicists work in probabilities instead.

## The Heisenberg uncertainty principle

The fact that matter exhibits wave-like properties gives rise to more trouble — waves aren't localized in space. And knowing that inspired Werner Heisenberg, in 1927, to come up with his celebrated uncertainty principle.

You can completely describe objects in classical physics by their momentum and position, both of which you can measure exactly. In other words, classical physics is completely *deterministic*.

On the atomic level, however, quantum physics paints a different picture. Here, the *Heisenberg uncertainty principle* says that there's an inherent uncertainty in the relation between position and momentum. In the $x$ direction, for example, that looks like this:

$$\Delta x \Delta p_x \geq \frac{\hbar}{2}$$

where $\Delta x$ is the measurement uncertainty in the particle's $x$ position, $\delta p_x$ is its measurement uncertainty in its momentum in the $x$ direction and $\hbar = h/2\pi$.

That is to say, the more accurately you know the position of a particle, the less accurately you know the momentum, and vice versa. This relation holds for all three dimensions:

$$\Delta y \Delta p_y \geq \frac{\hbar}{2}$$

$$\Delta z \Delta p_z \geq \frac{\hbar}{2}$$

And the Heisenberg uncertainty principle is a direct consequence of the wave-like nature of matter, because you can't completely pin down a wave.

Quantum physics, unlike classical physics, is completely undeterministic. You can never know the *precise* position and momentum of a particle at any one time. You can give only probabilities for these linked measurements.

## Rolling the dice: Quantum physics and probability

In quantum physics, the state of a particle is described by a wave function, $\psi(r, t)$. The wave function describes the de Broglie wave of a particle, giving its amplitude as a function of position and time. (See the earlier section "A Dual Identity: Looking at Particles as Waves" for more on de Broglie.)

Note that the wave function gives a particle's amplitude, not intensity; if you want to find the intensity of the wave function, you have to square it: $|\psi(r, t)|^2$. The *intensity* of a wave is what's equal to the probability that the particle will be at that position at that time.

That's how quantum physics converts issues of momentum and position into probabilities: by using a wave function, whose square tells you the *probability density* that a particle will occupy a particular position or have a particular momentum. In other words, $|\psi(r, t)|^2 d^3r$ is the probability that the particle will be found in the volume element $d^3r$, located at position $r$ at time $t$.

Besides the position-space wave function $\psi(r, t)$, there's also a momentum-space version of the wave function: $\phi(p, t)$.

This book is largely a study of the wave function — the wave functions of free particles, the wave functions of particles trapped inside potentials, of identical particles hitting each other, of particles in harmonic oscillation, of light scattering from particles, and more. Using this kind of physics, you can predict the behavior of all kinds of physical systems.

# Chapter 2

# Entering the Matrix: Welcome to State Vectors

. . . . . . . . . . . . . . . . . . . . . . . . . . . . . . . . . . . . . . . . . . . . . . .

*In This Chapter*

▶ Creating state vectors

▶ Using Dirac notation for state vectors

▶ Working with bras and kets

▶ Understanding matrix mechanics

▶ Getting to wave mechanics

. . . . . . . . . . . . . . . . . . . . . . . . . . . . . . . . . . . . . . . . . . . . . . .

*Q*uantum physics isn't just about playing around with your particle accelerator while trying not to destroy the universe. Sometimes, you get to do things that are a little more mundane, like turn lights off and on, perform a bit of calculus, or play with dice.

If you're actually doing physics with those dice (beyond hurling them across the room), the lab director won't even get mad at you. In quantum physics, absolute measurements are replaced by probabilities, so you may use dice to calculate the probabilities that various numbers will come up. You can then assemble those values into a vector (single-column matrix) in Hilbert space (a type of infinitely dimensional vector space with some properties that are especially valuable in quantum physics).

This chapter introduces how you deal with probabilities in quantum physics, starting by viewing the various possible states a particle can occupy as a vector — a vector of probability states. From there, I help you familiarize yourself with some mathematical notations common in quantum physics, including bras, kets, matrices, and wave functions. Along the way, you also get to work with some important operators.

# Creating Your Own Vectors in Hilbert Space

In quantum physics, probabilities take the place of absolute measurements. Say you've been experimenting with rolling a pair of dice and are trying to figure the relative probability that the dice will show various values. You come up with a list indicating the relative probability of rolling a 2, 3, 4, and so on, all the way up to 12:

| Sum of the Dice | Relative Probability (Number of Ways of Rolling a Particular Total) |
|---|---|
| 2 | 1 |
| 3 | 2 |
| 4 | 3 |
| 5 | 4 |
| 6 | 5 |
| 7 | 6 |
| 8 | 5 |
| 9 | 4 |
| 10 | 3 |
| 11 | 2 |
| 12 | 1 |

In other words, you're twice as likely to roll a 3 than a 2, you're four times as likely to roll a 5 than a 2, and so on. You can assemble these relative probabilities into a vector (if you're thinking of a "vector" from physics, think in terms of a column of the vector's components, not a magnitude and direction) to keep track of them easily:

$$\begin{vmatrix} 1 \\ 2 \\ 3 \\ 4 \\ 5 \\ 6 \\ 5 \\ 4 \\ 3 \\ 2 \\ 1 \end{vmatrix}$$

Okay, now you're getting closer to the way quantum physics works. You have a vector of the probabilities that the dice will occupy various states. However, quantum physics doesn't deal directly with probabilities but rather with *probability amplitudes,* which are the square roots of the probabilities. To find the actual probability that a particle will be in a certain state, you add wave functions — which are going to be represented by these vectors — and then square them (see Chapter 1 for info on why). So take the square root of all these entries to get the probability amplitudes:

$$\begin{vmatrix} 1 \\ 2^{1/2} \\ 3^{1/2} \\ 2 \\ 5^{1/2} \\ 6^{1/2} \\ 5^{1/2} \\ 2 \\ 3^{1/2} \\ 2^{1/2} \\ 1 \end{vmatrix}$$

That's better, but adding the squares of all these should add up to a total probability of 1; as it is now, the sum of the squares of these numbers is 36, so divide each entry by $36^{1/2}$, or 6:

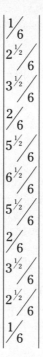

So now you can get the probability of rolling any combination from 2 to 12 by reading down the vector — the probability of rolling a 2 is $\frac{1}{6}$, of rolling a 3 is $\frac{\sqrt{2}}{6}$, and so on.

# Making Life Easier with Dirac Notation

When you have a state vector that gives the probability amplitude that a pair of dice will be in their various possible states, you basically have a vector in *dice space* — all the possible states that a pair of dice can take, which is an 11-dimensional space. (See the preceding section for more on state vectors.)

But in most quantum physics problems, the vectors can be infinitely large — for example, a moving particle can be in an infinite number of states. Handling large arrays of states isn't easy using vector notation, so instead of explicitly writing out the whole vector each time, quantum physics usually uses the notation developed by physicist Paul Dirac — the *Dirac* or *bra-ket notation*.

## Abbreviating state vectors as kets

Dirac notation abbreviates the state vector as a *ket*, like this: $| \psi >$. So in the dice example, you can write the state vector as a ket this way:

$| \psi > =$

$$\begin{vmatrix} {}^{1}\!/_{6} \\ 2^{1/2}\!/_{6} \\ 3^{1/2}\!/_{6} \\ {}^{2}\!/_{6} \\ 5^{1/2}\!/_{6} \\ 6^{1/2}\!/_{6} \\ 5^{1/2}\!/_{6} \\ {}^{2}\!/_{6} \\ 3^{1/2}\!/_{6} \\ 2^{1/2}\!/_{6} \\ {}^{1}\!/_{6} \end{vmatrix}$$

Here, the components of the state vector are represented by numbers in 11-dimensional dice space. More commonly, however, each component represents a function, something like this:

$$|\psi> =
\begin{vmatrix}
\dfrac{1}{6} e^{ikx-\omega t} \\[4pt]
\dfrac{2^{1/2}}{6} e^{i2kx-\omega t} \\[4pt]
\dfrac{3^{1/2}}{6} e^{i3kx-\omega t} \\[4pt]
\dfrac{2}{6} e^{i4kx-\omega t} \\[4pt]
\dfrac{5^{1/2}}{6} e^{i5kx-\omega t} \\[4pt]
\dfrac{6^{1/2}}{6} e^{i6kx-\omega t} \\[4pt]
\dfrac{5^{1/2}}{6} e^{i7kx-\omega t} \\[4pt]
\dfrac{2}{6} e^{i8kx-\omega t} \\[4pt]
\dfrac{3^{1/2}}{6} e^{i9kx-\omega t} \\[4pt]
\dfrac{2^{1/2}}{6} e^{i10kx-\omega t} \\[4pt]
\dfrac{1}{6} e^{i11kx-\omega t}
\end{vmatrix}$$

You can use functions as components of a state vector as long as they're linearly independent functions (and so can be treated as independent axes in Hilbert space). In general, a set of vectors $\phi_N$ in Hilbert space is linearly independent if the only solution to the following equation is that all the coefficients $a_n = 0$:

$$\sum_{i=1}^{N} a_n \phi_i = 0$$

That is, as long as you can't write any one vector as a linear combination of the others, the vectors are linearly independent and so form a valid basis in Hilbert space.

## Writing the Hermitian conjugate as a bra

For every ket, there's a corresponding *bra*. (The terms come from *bra-ket,* or *bracket,* which should be clearer in the upcoming section titled "Grooving with Operators.") A *bra* is the Hermitian conjugate of the corresponding ket.

Suppose you start with this ket:

$$|\psi> = \begin{vmatrix} 1/6 \\ 2^{1/2}/6 \\ 3^{1/2}/6 \\ 2/6 \\ 5^{1/2}/6 \\ 6^{1/2}/6 \\ 5^{1/2}/6 \\ 2/6 \\ 3^{1/2}/6 \\ 2^{1/2}/6 \\ 1/6 \end{vmatrix}$$

The $^*$ symbol means the complex conjugate. (A *complex conjugate* flips the sign connecting the real and imaginary parts of a complex number.) So the corresponding bra, which you write as $<\psi|$, equals $|\psi>^{T*}$. The bra is this row vector:

$$<\psi| = \begin{vmatrix} 1/6 & 2^{1/2}/6 & 3^{1/2}/6 & 2/6 & 5^{1/2}/6 & 6^{1/2}/6 & 5^{1/2}/6 & 2/6 & 3^{1/2}/6 & 2^{1/2}/6 & 1/6 \end{vmatrix}$$

Note that if any of the elements of the ket are complex numbers, you have to take their complex conjugate when creating the associated bra. For instance, if your complex number in the ket is $a + bi,$ its complex conjugate in the bra is $a - bi.$

## Multiplying bras and kets: A probability of 1

You can take the product of your ket and bra, denoted as $\langle\psi|\psi\rangle$, like this:

$$\langle\psi|\psi\rangle =$$

$$\left|\frac{1}{6}\ \frac{2^{\frac{1}{2}}}{6}\ \frac{3^{\frac{1}{2}}}{6}\ \frac{2}{6}\ \frac{5^{\frac{1}{2}}}{6}\ \frac{6^{\frac{1}{2}}}{6}\ \frac{5^{\frac{1}{2}}}{6}\ \frac{2}{6}\ \frac{3^{\frac{1}{2}}}{6}\ \frac{2^{\frac{1}{2}}}{6}\ \frac{1}{6}\right| \left|\begin{array}{c}\frac{1}{6}\\[4pt]\frac{2^{\frac{1}{2}}}{6}\\[4pt]\frac{3^{\frac{1}{2}}}{6}\\[4pt]\frac{2}{6}\\[4pt]\frac{5^{\frac{1}{2}}}{6}\\[4pt]\frac{6^{\frac{1}{2}}}{6}\\[4pt]\frac{5^{\frac{1}{2}}}{6}\\[4pt]\frac{2}{6}\\[4pt]\frac{3^{\frac{1}{2}}}{6}\\[4pt]\frac{2^{\frac{1}{2}}}{6}\\[4pt]\frac{1}{6}\end{array}\right|$$

This is just matrix multiplication, and the result is the same as taking the sum of the squares of the elements:

$$\langle\psi|\psi\rangle = \frac{1}{36}+\frac{2}{36}+\frac{3}{36}+\frac{4}{36}+\frac{5}{36}+\frac{6}{36}+\frac{5}{36}+\frac{4}{36}+\frac{3}{36}+\frac{2}{36}+\frac{1}{36}=1$$

And that's the way it should be, because the total probability should add up to 1. Therefore, in general, the product of the bra and ket equals 1:

$$\langle\psi|\psi\rangle = 1$$

If this relation holds, the ket $|\psi\rangle$ is said to be *normalized*.

# Covering all your bases: Bras and kets as basis-less state vectors

The reason ket notation, $|\psi>$, is so popular in quantum physics is that it allows you to work with state vectors in a basis-free way. In other words, you're not stuck in the position basis, the momentum basis, or the energy basis. That's helpful, because most of the work in quantum physics takes place in abstract calculations, and you don't want to have to drag all the components of state vectors through those calculations (often you can't — there may be infinite possible states in the problem you're dealing with).

For example, say that you're representing your states using position vectors in a three-dimensional Hilbert space — that is, you have $x$, $y$, and $z$ axes, forming a position *basis* for your space. That's fine, but not all your calculations have to be done using that position basis.

You may want to, for example, represent your states in a three-dimensional momentum space, with three axes in Hilbert space, $p_x$, $p_y$, and $p_z$. Now you'd have to change all your position vectors to momentum vectors, adjusting each component, and keep track of what happens to every component through all your calculations.

So Dirac's bra-ket notation comes to the rescue here — you use it to perform all the math and then plug in the various components of your state vectors as needed at the end. That is, you can perform your calculations in purely symbolic terms, without being tied to a basis.

And when you need to deal with the components of a ket, such as when you want to get physical answers, you can also convert kets to a different basis by taking the ket's components along the axes of that basis. For example, if you want to convert the ket $|\psi>$ to the position basis, as represented by *i, j*, and *k*, which are position-unit vectors along the $x$, $y$, and $z$ axes, you can just find the three components of $|\psi>$ along *i, j*, and *k* for the new version of the ket, $|\phi>$. Here's how that looks in general, where $\phi_i$ are unit vectors in the basis you're switching to:

$$|\phi> = \sum_{i=1}^{N} < \psi|\phi_i >$$

# Understanding some relationships using kets

Ket notation makes the math easier than it is in matrix form because you can take advantage of a few mathematical relationships. For example, here's the so-called Schwarz inequality for state vectors:

$$\left|<\psi|\phi>\right|^2 \le <\psi|\psi><\phi|\phi>$$

This says that the square of the absolute value of the product of two state vectors, $|<\psi|\phi>|^2$, is less than or equal to $<\psi|\psi><\phi|\phi>$. This turns out the be the analog of the vector inequality:

$$\left|A \cdot B\right|^2 \le \left|A\right|^2 \left|B\right|^2$$

So why is the Schwarz inequality so useful? It turns out that you can derive the Heisenberg uncertainty principle from it (see Chapter 1 for more on this principle).

Other ket relationships can also simplify your calculations. For instance, two kets, $|\psi>$ and $|\phi>$, are said to be *orthogonal* if

$$<\psi|\phi>=0$$

And two kets are said to be *orthonormal* if they meet the following conditions:

- $<\psi|\phi>=0$
- $<\psi|\psi>=1$
- $<\phi|\phi>=1$

With this information in mind, you're now ready to start working with operators.

# Grooving with Operators

What about all the calculations that you're supposed to be able to perform with kets? Taking the product of a bra and a ket, $<\psi|\phi>$, is fine as far as it goes, but what about extracting some physical quantities you can measure? That's where operators come in.

## Hello, operator: How operators work

Here's the general definition of an operator A in quantum physics: An *operator* is a mathematical rule that, when operating on a ket, $|\psi>$, transforms that ket into a new ket, $|\psi'>$ in the same space (which could just be the old ket multiplied by a scalar). So when you have an operator A, it transforms kets like this:

$$A|\psi> = |\psi'>$$

For that matter, the same operator can also transform bras:

$$<\psi|A = <\psi'|$$

Here are several examples of the kinds of operators you'll see:

- **Hamiltonian (H):** Applying the Hamiltonian operator (which looks different for every different physical situation) gives you E, the energy of the particle represented by the ket |ψ>; E is a scalar quantity:

$$H|\psi> = E|\psi>$$

- **Unity or identity (I):** The unity or identity operator leaves kets unchanged:

$$I|\psi> = |\psi>$$

- **Gradient (∇):** The gradient operator works like this:

$$\nabla|\psi> = \frac{\partial}{\partial x}|\psi>\boldsymbol{i} + \frac{\partial}{\partial y}|\psi>\boldsymbol{j} + \frac{\partial}{\partial z}|\psi>\boldsymbol{k}$$

- **Linear momentum (P):** The linear momentum operator looks like this in quantum mechanics:

$$P|\psi> = -i\hbar\nabla|\psi>$$

- **Laplacian (Δ):** You use the Laplacian operator, which is much like a second-order gradient, to create the energy-finding Hamiltonian operator:

$$\Delta^2|\psi> = \Delta|\psi> = \frac{\partial^2}{\partial x^2}|\psi> + \frac{\partial^2}{\partial y^2}|\psi> + \frac{\partial^2}{\partial z^2}|\psi>$$

In general, multiplying operators together is not the same independent of order, so for the operators A and B, AB ≠ BA.

And an operator A is said to be *linear* if

$$A\left(c_1\big|\psi\!>+c_2\big|\psi\!>\right)=c_1 A\big|\psi\!>+c_2 A\big|\psi\!>$$

# I expected that: Finding expectation values

Given that everything in quantum physics is done in terms of probabilities, making predictions becomes very important. And the biggest such prediction is the expectation value. The *expectation value* of an operator is the average value that you would measure if you performed the measurement many times. For example, the expectation value of the Hamiltonian operator (see the preceding section) is the average energy of the system you're studying.

The expectation value is a weighted average of the probabilities of the system's being in its various possible states. Here's how you find the expectation value of an operator A:

$$\text{Expectation value } =<\psi\big|A\big|\psi>$$

Note that because you can express <ψ| as a row operator and |ψ> as a column vector, you can express the operator A as a square matrix.

For example, suppose you're working with a pair of dice and the probabilities of all the possible sums (see the earlier section "Creating Your Own Vectors in Hilbert Space"). In this dice example, the expectation value is a sum of terms, and each term is a value that can be displayed by the dice, multiplied by the probability that that value will appear.

The bra and ket will handle the probabilities, so it's up to the operator that you create for this — call it the *Roll operator,* R — to store the dice values (2 through 12) for each probability. Therefore, the operator R looks like this:

$R =$

$$\begin{vmatrix} 200 & 0 & 0 & 0 & 0 & 0 & 0 & 0 & 0 & 0 \\ 0 & 300 & 0 & 0 & 0 & 0 & 0 & 0 & 0 & 0 \\ 0 & 0 & 400 & 0 & 0 & 0 & 0 & 0 & 0 & 0 \\ 0 & 0 & 0 & 500 & 0 & 0 & 0 & 0 & 0 & 0 \\ 0 & 0 & 0 & 0 & 600 & 0 & 0 & 0 & 0 & 0 \\ 0 & 0 & 0 & 0 & 0 & 700 & 0 & 0 & 0 & 0 \\ 0 & 0 & 0 & 0 & 0 & 0 & 800 & 0 & 0 & 0 \\ 0 & 0 & 0 & 0 & 0 & 0 & 0 & 9. & 0 & 0 & 0 \\ 0 & 0 & 0 & 0 & 0 & 0 & 0 & 0 & 10 & 0 & 0 \\ 0 & 0 & 0 & 0 & 0 & 0 & 0 & 0 & 0 & 11 & 0 \\ 0 & 0 & 0 & 0 & 0 & 0 & 0 & 0 & 0 & 0 & 12 \end{vmatrix}$$

So to find the expectation value of R, you need to calculate $\langle\psi|R|\psi\rangle$. Spelling that out in terms of components gives you the following:

$\langle\psi|R|\psi\rangle =$

$$\begin{vmatrix} \tfrac{1}{6} & \tfrac{2^{1/2}}{6} & \tfrac{3^{1/2}}{6} & \tfrac{2}{6} & \tfrac{5^{1/2}}{6} & \tfrac{6^{1/2}}{6} & \tfrac{5^{1/2}}{6} & \tfrac{2}{6} & \tfrac{3^{1/2}}{6} & \tfrac{2^{1/2}}{6} & \tfrac{1}{6} \end{vmatrix} \begin{Vmatrix} 2 & 0 & 0 & 0 & 0 & 0 & 0 & 0 & 0 & 0 & 0 \\ 0 & 3 & 0 & 0 & 0 & 0 & 0 & 0 & 0 & 0 & 0 \\ 0 & 0 & 4 & 0 & 0 & 0 & 0 & 0 & 0 & 0 & 0 \\ 0 & 0 & 0 & 5 & 0 & 0 & 0 & 0 & 0 & 0 & 0 \\ 0 & 0 & 0 & 0 & 6 & 0 & 0 & 0 & 0 & 0 & 0 \\ 0 & 0 & 0 & 0 & 0 & 7 & 0 & 0 & 0 & 0 & 0 \\ 0 & 0 & 0 & 0 & 0 & 0 & 8 & 0 & 0 & 0 & 0 \\ 0 & 0 & 0 & 0 & 0 & 0 & 0 & 9 & 0 & 0 & 0 \\ 0 & 0 & 0 & 0 & 0 & 0 & 0 & 0 & 10 & 0 & 0 \\ 0 & 0 & 0 & 0 & 0 & 0 & 0 & 0 & 0 & 11 & 0 \\ 0 & 0 & 0 & 0 & 0 & 0 & 0 & 0 & 0 & 0 & 12 \end{Vmatrix} \begin{Vmatrix} \tfrac{1}{6} \\ \tfrac{2^{1/2}}{6} \\ \tfrac{3^{1/2}}{6} \\ \tfrac{2}{6} \\ \tfrac{5^{1/2}}{6} \\ \tfrac{6^{1/2}}{6} \\ \tfrac{5^{1/2}}{6} \\ \tfrac{2}{6} \\ \tfrac{3^{1/2}}{6} \\ \tfrac{2^{1/2}}{6} \\ \tfrac{1}{6} \end{Vmatrix}$$

Doing the math, you get

$\langle\psi|R|\psi\rangle = 7$

So the expectation value of a roll of the dice is 7. Now you can see where the terms *bra* and *ket* come from — they "bracket" an operator to give you expectation values. In fact, the expectation value is such a common thing to find that you'll often find <ψ|R|ψ> abbreviated as <R>, so

<R> = 7

## Looking at linear operators

An operator A is said to be *linear* if it meets the following condition:

$A(c_1|ψ> + c_2|ψ>) = c_1A|ψ> + c_2A|ψ>$

For instance, the expression |φ><ψ| is actually a linear operator. To see that, apply |φ><ψ| to a ket, |χ>:

$|φ><ψ|χ>$

You can also write this as

$<ψ|χ>|φ>$

The expression <ψ|χ> is always a complex number (which could be purely real), so this breaks down to

$c|φ>$

where *c* is a complex number. Thus, |φ><ψ| is indeed a linear operator.

# Going Hermitian with Hermitian Operators and Adjoints

The *Hermitian adjoint* — also called the *adjoint* or *Hermitian conjugate* — of an operator A is denoted A†. To find the hermitian adjoint, follow these steps:

1. **Replace complex constants with their complex conjugates.**

   The Hermitian adjoint of a complex number is the complex conjugate of that number:

   $a^† = a^*$

2. **Replace kets with their corresponding bras, and replace bras with their corresponding kets.**

You have to exchange the bras and kets when finding the Hermitian adjoint of an operator, so finding the Hermitian adjoint of an operator is not just the same as mathematically finding its complex conjugate.

3. **Replace operators with their Hermitian operators.**

In quantum mechanics, operators that are equal to their Hermitian adjoints are called *Hermitian operators*. In other words, an operator is Hermitian if

$$A^\dagger = A$$

Hermitian operators appear throughout the book, and they have special properties. For instance, the matrix that represents them may be *diagonalized* — that is, written so that the only nonzero elements appear along the matrix's diagonal. Also, the expectation value of a Hermitian operator is guaranteed to be a real number, not complex (see the earlier section "I expected that: Finding expectation values").

4. **Write your final equation.**

$$< \psi | A^\dagger | \phi > = < \phi | A | \psi$$

Here are some relationships concerning Hermitian adjoints:

- $\left(aA\right)^\dagger = a^* A^\dagger$
- $\left(A^\dagger\right)^\dagger = A$
- $\left(A+B\right)^\dagger = A^\dagger + B^\dagger$
- $\left(AB\right)^\dagger = B^\dagger A^\dagger$
- $\left(AB|\psi>\right)^\dagger = < \psi | B^\dagger A^\dagger$

# Forward and Backward: Finding the Commutator

The measure of how different it is to apply operator A and then B, versus B and then A, is called the operators' *commutator*. Here's how you define the commutator of operators A and B:

$$[A, B] = AB - BA$$

## Commuting

Two operators *commute* with each other if their commutator is equal to zero. That is, it doesn't make any difference in what order you apply them:

[A, B] = 0

Note in particular that any operator commutes with itself:

[A, A] = 0

And it's easy to show that the commutator of A, B is the negative of the commutator of B, A:

[A, B] = –[B, A]

It's also true that commutators are linear — that is, $A(c_1|\psi> + c_2|\psi>) = c_1A|\psi> + c_2A|\psi>$:

[A, B + C + D + ...] = [A, B] + [A, C] + [A, D] + ...

And the Hermitian adjoint of a commutator works this way:

$$\left[A, B\right]^{\dagger} = \left[B^{\dagger}, A^{\dagger}\right]$$

You can also find the anticommutator, {A, B}:

{A, B} = AB + BA

## Finding anti-Hermitian operators

Here's another one: What can you say about the Hermitian adjoint of the commutator of two Hermitian operators? Here's the answer. First, write the adjoint:

$$\left[A, B\right]^{\dagger}$$

The definition of commutators tells you the following:

$$\left[A, B\right]^{\dagger} = \left(AB - BA\right)^{\dagger}$$

You know $(AB)^{\dagger} = B^{\dagger}A^{\dagger}$ (see the earlier section "Going Hermitian with Hermitian Operators and Adjoints" for properties of adjoints). Therefore,

$$\left[A, B\right]^{\dagger} = \left(AB - BA\right)^{\dagger} = B^{\dagger}A^{\dagger} - A^{\dagger}B^{\dagger}$$

But for Hermitian operators, $A = A^\dagger$, so remove the $^\dagger$ symbols:

$$\left[A, B\right]^\dagger = \left(AB - BA\right)^\dagger = B^\dagger A^\dagger - A^\dagger B^\dagger = BA - AB$$

But BA – AB is just –[A, B], so you have the following:

$$\left[A, B\right]^\dagger = -\left[A, B\right]$$

A and B here are Hermitian operators. When you take the Hermitian adjoint of an expression and get the same thing back with a negative sign in front of it, the expression is called *anti-Hermitian,* so the commutator of two Hermitian operators is anti-Hermitian. (And by the way, the expectation value of an anti-Hermitian operator is guaranteed to be completely imaginary.)

# *Starting from Scratch and Ending Up with Heisenberg*

If you've read through the last few sections, you're now armed with all this new technology: Hermitian operators and commutators. How can you put it to work? You can come up with the Heisenberg uncertainty relation starting virtually from scratch.

Here's a calculation that takes you from a few basic definitions to the Heisenberg uncertainty relation. This kind of calculation shows how much easier it is to use the basis-less bra and ket notation than the full matrix version of state vectors. This isn't the kind of calculation that you'll need to do in class, but follow it through — knowing how to use kets, bras, commutators, and Hermitian operators is vital in the coming chapters.

The uncertainty in a measurement of the Hermitian operator named A is formally given by

$$\Delta A = \left(< A^2 > - < A >^2\right)^{1/2}$$

That is, $\Delta A$ is equal to the square root of the expectation value of $A^2$ minus the squared expectation value of A. If you've taken any math classes that dealt with statistics, this formula may be familiar to you. Similarly, the uncertainty in a measurement using Hermitian operator B is

$$\Delta B = \left(< B^2 > - < B >^2\right)^{1/2}$$

Now consider the *operators* $\Delta A$ and $\Delta B$ (not the uncertainties $\Delta A$ and $\Delta B$ anymore), and assume that applying $\Delta A$ and $\Delta B$ as operators gives you measurement values like this:

$$\Delta A = A - <A>$$

$$\Delta B = B - <B>$$

Like any operator, using $\Delta A$ and $\Delta B$ can result in new kets:

$$\Delta A |\psi> = |\chi>$$

$$\Delta B |\psi> = |\phi>$$

Here's the key: The Schwarz inequaility (from the earlier section "Understanding some relationships using kets") gives you

$$<\chi|\chi><\phi|\phi> \geq \left|<\psi|\phi>\right|^2$$

So you can see that the inequality sign, $\geq$, which plays a big part in the Heisenberg uncertainty relation, has already crept into the calculation.

Because $\Delta A$ and $\Delta B$ are Hermitian, $<\chi|\chi>$ is equal to $<\psi|\Delta A^2|\psi>$ and $<\phi|\phi>$ is equal to $<\psi|\Delta B^2|\psi>$. Because $\Delta A^\dagger = \Delta A$ (the definition of a Hermitian operator), you can see that

$$<\chi|\chi> = <\psi|\Delta A^\dagger \Delta A|\psi>$$

This means that

$$<\chi|\chi> = <\psi|\Delta A^\dagger \Delta A|\psi> = <\psi \Delta A^2|\psi>$$

That is, $<\chi|\chi>$ is equal to $<\Delta A^2>$ and $<\phi|\phi>$ is equal to $<\Delta B^2>$. So you can rewrite the Schwarz inequality like this:

$$<\Delta A^2><\Delta B^2> \geq \left|<\Delta A \Delta B>\right|^2$$

Okay, where has this gotten you? It's time to be clever. Note that you can write $\Delta A \Delta B$ as

$$\Delta A \Delta B = \frac{1}{2}\left[\Delta A,\ \Delta B\right] + \frac{1}{2}\left\{\Delta A,\ \Delta B\right\}$$

Here, $\{\Delta A, \Delta B\} = \Delta A \Delta B + \Delta B \Delta A$ is the anticommutator of the operators $\Delta A$ and $\Delta B$. Because $[\Delta A, \Delta B] = [A, B]$ (the constants <A> and <B> subtract out), you can rewrite this equation:

$$\Delta A \Delta B = \frac{1}{2}\left[A, B\right] + \frac{1}{2}\{\Delta A, \Delta B\}$$

Here's where the math gets intense. Take a look at what you know so far:

✔ The commutator of two Hermitian operators, $[A, B]$, is anti-Hermitian.

✔ The expectation value of an anti-Hermitian is imaginary.

✔ $\{\Delta A, \Delta B\}$ is Hermitian.

✔ The expectation value of a Hermitian is real.

All this means that you can view the expectation value of the equation as the sum of real ($\{\Delta A, \Delta B\}$) and imaginary ($[A, B]$) parts, so

$$\left|< \Delta A \Delta B>\right|^2 = \frac{1}{4}\left|<\left[A, B\right]>\right|^2 + \frac{1}{4}\left|\{\Delta A, \Delta B\}\right|^2$$

And because the second term on the right is positive or zero, you can say that the following is true:

$$\left|< \Delta A \Delta B>\right|^2 \geq \frac{1}{4}\left|<\left[A, B\right]>\right|^2$$

Whew! But now compare this equation to the relationship from the earlier use of the Schwarz inequality:

$$< \Delta A^2 >< \Delta B^2 > \geq \left|< B\Delta A \Delta A >\right|^2$$

Combining the two equations gives you this:

$$< \Delta A^2 >< \Delta B^2 > \geq \frac{1}{4}\left|< A, B >\right|^2$$

This has the look of the Heisenberg uncertainty relation, except for the pesky expectation value brackets, < >, and the fact that $\Delta A$ and $\Delta B$ appear squared here. You want to reproduce the Heisenberg uncertainty relation here, which looks like this:

$$\Delta x \Delta p_x \geq \frac{\hbar}{2}$$

Okay, so how do you get the left side of the equation from $<\Delta A^2><\Delta B^2>$ to $\Delta A\Delta B$? Because an earlier equation tells you that $\Delta A = A - <A>$, you know the following:

$$< \Delta A^2 > = < A^2 + < A >^2 - 2A<A>>$$

Taking the expectation value of the last term in this equation, you get this result:

$$< \Delta A^2 > = < A^2 + < A >^2 - 2A<A>> = < A^2 > - < A >^2$$

Square the earlier equation $\Delta A = (<A^2> - <A>^2)^{1/2}$ to get the following:

$$\Delta A^2 = < A^2 > - < A >^2$$

And comparing that equation to the before it, you conclude that

$$< \Delta A^2 > = \Delta A^2$$

Cool. That result means that $: \Delta A^2 > < \Delta B^2 > \geq \frac{1}{4}\left|<\left[A, B\right]>\right|^2$ becomes

$$\Delta A^2 \Delta B^2 \geq \frac{1}{4}\left|<\left[A, B\right]>\right|^2$$

This inequality at last means that

$$\Delta A\Delta B \geq \frac{1}{2}\left|<\left[A, B\right]>\right|$$

Well, well, well. So the product of two uncertainties is greater than or equal to $1/2$ the absolute value of the commutator of their respective operators? Wow. Is that the Heisenberg uncertainty relation? Well, take a look. In quantum mechanics, the momentum operator looks like this:

$$P = -i\hbar\nabla$$

And the operator for the momentum in the $x$ direction is

$$P_x = -i\hbar\frac{\partial}{\partial x}$$

So what's the commutator of the X operator (which just returns the $x$ position of a particle) and $P_x$? $[X, P_x] = -i\hbar$, so from $\Delta A\Delta B \geq \frac{1}{2}\left|<\left[A, B\right]>\right|$, you get this next equation (remember, $\Delta x$ and $\Delta p_x$ here are the uncertainties in $x$ and $\Delta p_x$, not the operators):

$$\Delta x \Delta p_x \geq \frac{h}{2}$$

Hot dog! That is the Heisenberg uncertainty relation. (Notice that by deriving it from scratch, however, you haven't actually constrained the physical world through the use of abstract mathematics — you've merely proved, using a few basic assumptions, that you can't *measure* the physical world with perfect accuracy.)

# Eigenvectors and Eigenvalues: They're Naturally Eigentastic!

As you know if you've been following along in this chapter, applying an operator to a ket can result in a new ket:

$$A|\psi> = |\chi>$$

To make things easier, you can work with eigenvectors and eigenvalues (*eigen* is German for "innate" or "natural"). For example, $|\psi>$ is an *eigenvector* of the operator A if

- The number $a$ is a complex constant
- $A|\psi> = a|\psi>$

Note what's happening here: Applying A to one of its eigenvectors, $|\psi>$, gives you $|\psi>$ back, multiplied by that eigenvector's *eigenvalue, a*.

Although $a$ can be a complex constant, the eigenvalues of Hermitian operators are real numbers, and their eigenvectors are orthogonal (that is, $<\psi|\phi> = 0$).

Casting a problem in terms of eigenvectors and eigenvalues can make life a lot easier because applying the operator to its eigenvectors merely gives you the same eigenvector back again, multiplied by its eigenvalue — there's no pesky change of state, so you don't have to deal with a different state vector.

Take a look at this idea, using the R operator from rolling the dice, which is expressed this way in matrix form (see the earlier section "I expected that: Finding expectation values" for more on this matrix):

$R =$

$$
\begin{vmatrix}
2 & 0 & 0 & 0 & 0 & 0 & 0 & 0 & 0 & 0 & 0 \\
0 & 3 & 0 & 0 & 0 & 0 & 0 & 0 & 0 & 0 & 0 \\
0 & 0 & 4 & 0 & 0 & 0 & 0 & 0 & 0 & 0 & 0 \\
0 & 0 & 0 & 5 & 0 & 0 & 0 & 0 & 0 & 0 & 0 \\
0 & 0 & 0 & 0 & 6 & 0 & 0 & 0 & 0 & 0 & 0 \\
0 & 0 & 0 & 0 & 0 & 7 & 0 & 0 & 0 & 0 & 0 \\
0 & 0 & 0 & 0 & 0 & 0 & 8 & 0 & 0 & 0 & 0 \\
0 & 0 & 0 & 0 & 0 & 0 & 0 & 9 & 0 & 0 & 0 \\
0 & 0 & 0 & 0 & 0 & 0 & 0 & 0 & 10 & 0 & 0 \\
0 & 0 & 0 & 0 & 0 & 0 & 0 & 0 & 0 & 11 & 0 \\
0 & 0 & 0 & 0 & 0 & 0 & 0 & 0 & 0 & 0 & 12
\end{vmatrix}
$$

The R operator works in 11-dimensional space and is Hermitian, so there'll be 11 orthogonal eigenvectors and 11 corresponding eigenvalues.

Because R is a diagonal matrix, finding the eigenvectors is easy. You can take unit vectors in the eleven different directions as the eigenvectors. Here's what the first eigenvector, $\xi_1$, would look like:

$\xi_1 =$

$$
\begin{vmatrix}
1 \\
0 \\
0 \\
0 \\
0 \\
0 \\
0 \\
0 \\
0 \\
0 \\
0
\end{vmatrix}
$$

And here's what the second eigenvector, $\xi_2$, would look like:

$$\xi_2 = \begin{vmatrix} 0 \\ 1 \\ 0 \\ 0 \\ 0 \\ 0 \\ 0 \\ 0 \\ 0 \\ 0 \\ 0 \end{vmatrix}$$

And so on, up to $\xi_{11}$:

$$\xi_{11} = \begin{vmatrix} 0 \\ 0 \\ 0 \\ 0 \\ 0 \\ 0 \\ 0 \\ 0 \\ 0 \\ 0 \\ 1 \end{vmatrix}$$

Note that all the eigenvectors are orthogonal.

And the eigenvalues? They're the numbers you get when you apply the R operator to an eigenvector. Because the eigenvectors are just unit vectors in all 11 dimensions, the eigenvalues are the numbers on the diagonal of the R matrix: 2, 3, 4, and so on, up to 12.

## *Understanding how they work*

The eigenvectors of a Hermitian operator define a complete set of orthonormal vectors — that is, a complete basis for the state space. When viewed in this "eigenbasis," which is built of the eigenvectors, the operator in matrix format is diagonal and the elements along the diagonal of the matrix are the eigenvalues.

This arrangement is one of the main reasons working with eigenvectors is so useful; your original operator may have looked something like this (*Note:* Bear in mind that the elements in an operator can also be functions, not just numbers):

$$R =
\begin{vmatrix}
0 & 0 & 1 & 0 & 6 & 0 & 0 & 3 & 0 & 0 & 00 \\
0 & 1 & 0 & 0 & 0 & 9 & 0 & 0 & 0 & 0 & 0 & 0 \\
4 & 0 & 0 & 0 & 120 & 0 & 0 & 2 & 0 & 0 & 0 \\
9 & 0 & 0 & 0 & 0 & 0 & 7 & 0 & 0 & 0 & 0 \\
0 & 0 & 6 & 0 & 6 & 0 & 0 & 0 & 8 & 0 & 0 \\
1 & 0 & 3 & 0 & 5 & 0 & 0 & 0 & 0 & 0 & 8 \\
0 & 0 & 0 & 0 & 0 & 0 & 8 & 0 & 0 & 0 & 0 \\
0 & 0 & 0 & 0 & 0 & 0 & 0 & 9 & 0 & 0 & 0 \\
0 & 0 & 0 & 0 & 9 & 0 & 0 & 0 & 0 & 0 & 77 \\
0 & 0 & 0 & 0 & 110 & 0 & 0 & 0 & 1 & 0 \\
0 & 0 & 7 & 0 & 0 & 0 & 880 & 0 & 0 & 0 & 0
\end{vmatrix}$$

By switching to the basis of eigenvectors for the operator, you diagonalize the matrix into something more like what you've seen, which is much easier to work with:

$$R =
\begin{vmatrix}
20 & 0 & 0 & 0 & 0 & 0 & 0 & 0 & 0 & 0 & 0 \\
0 & 3 & 0 & 0 & 0 & 0 & 0 & 0 & 0 & 0 & 0 \\
0 & 0 & 4 & 0 & 0 & 0 & 0 & 0 & 0 & 0 & 0 \\
0 & 0 & 0 & 5 & 0 & 0 & 0 & 0 & 0 & 0 & 0 \\
0 & 0 & 0 & 0 & 6 & 0 & 0 & 0 & 0 & 0 & 0 \\
0 & 0 & 0 & 0 & 0 & 7 & 0 & 0 & 0 & 0 & 0 \\
0 & 0 & 0 & 0 & 0 & 0 & 8 & 0 & 0 & 0 & 0 \\
0 & 0 & 0 & 0 & 0 & 0 & 0 & 9 & 0 & 0 & 0 \\
0 & 0 & 0 & 0 & 0 & 0 & 0 & 0 & 100 & 0 & 0 \\
0 & 0 & 0 & 0 & 0 & 0 & 0 & 0 & 0 & 110 & 0 \\
0 & 0 & 0 & 0 & 0 & 0 & 0 & 0 & 0 & 0 & 12
\end{vmatrix}$$

You can see why the term *eigen* is applied to eigenvectors — they form a natural basis for the operator.

If two or more of the eigenvalues are the same, that eigenvalue is said to be *degenerate*. So for example, if three eigenvalues are equal to 6, then the eigenvalue 6 is threefold degenerate.

Here's another cool thing: If two Hermitian operators, A and B, commute, and if A doesn't have any degenerate eigenvalues, then each eigenvector of A is also an eigenvector of B. (See the earlier section "Forward and Backward: Finding the Commutator" for more on commuting.)

## Finding eigenvectors and eigenvalues

So given an operator in matrix form, how do you find its eigenvectors and eigenvalues? This is the equation you want to solve:

$$A|\psi> = a|\psi>$$

And you can rewrite this equation as the following:

$$(A - aI)|\psi> = 0$$

*I* represents the identity matrix, with 1s along its diagonal and 0s otherwise:

$$I =$$

$$\begin{vmatrix} 1 & 0 & 0 & 0 & 0 & 0 & 0 & 0 & 0 & 0... \\ 0 & 1 & 0 & 0 & 0 & 0 & 0 & 0 & 0 & 0... \\ 0 & 0 & 1 & 0 & 0 & 0 & 0 & 0 & 0 & 0... \\ 0 & 0 & 0 & 1 & 0 & 0 & 0 & 0 & 0 & 0... \\ 0 & 0 & 0 & 0 & 1 & 0 & 0 & 0 & 0 & 0... \\ 0 & 0 & 0 & 0 & 0 & 1 & 0 & 0 & 0 & 0... \\ & & & & \vdots & & & & & \end{vmatrix}$$

The solution to $(A - aI) |\psi> = 0$ exists only if the determinant of the matrix $A - aI$ is 0:

$$\det(A - aI) = 0$$

### Finding eigenvalues

Any values of $a$ that satisfy the equation $\det(A - aI) = 0$ are eigenvalues of the original equation. Try to find the eigenvalues and eigenvectors of the following matrix:

$$A = \begin{vmatrix} -1 & -1 \\ 2 & -4 \end{vmatrix}$$

First, convert the matrix into the form $A - aI$:

$$A - aI = \begin{vmatrix} -1-a & -1 \\ 2 & -4-a \end{vmatrix}$$

Next, find the determinant:

$$\det(A - aI) = (-1 - a)(-4 - a) + 2$$
$$\det(A - aI) = a^2 + 5a + 6$$

And this can be factored as follows:

$$\det(A - aI) = a^2 + 5a + 6 = (a + 2)(a + 3)$$

You know that $\det(A - aI) = 0$, so the eigenvalues of A are the roots of this equation; namely, $a_1 = -2$ and $a_2 = -3$.

### Finding eigenvectors

How about finding the eigenvectors? To find the eigenvector corresponding to $a_1$ (see the preceding section), substitute $a_1$ — the first eigenvalue, $-2$ — into the matrix in the form $A - aI$:

$$A - aI = \begin{vmatrix} -1-a & -1 \\ 2 & -4-a \end{vmatrix}$$

$$A - aI = \begin{vmatrix} 1 & -1 \\ 2 & -2 \end{vmatrix}$$

$$(A - aI)|\psi> = 0$$

So you have

$$\begin{vmatrix} 1 & -1 \\ 2 & -1 \end{vmatrix} \begin{vmatrix} \psi_1 \\ \psi_2 \end{vmatrix} = \begin{vmatrix} 0 \\ 0 \end{vmatrix}$$

Because every row of this matrix equation must be true, you know that $\psi_1 = \psi_2$. And that means that, up to an arbitrary constant, the eigenvector corresponding to $a_1$ is the following:

$$c\begin{vmatrix} 1 \\ 1 \end{vmatrix}$$

Drop the arbitrary constant, and just write this as a matrix:

$$\begin{vmatrix} 1 \\ 1 \end{vmatrix}$$

How about the eigenvector corresponding to $a_2$? Plugging $a_2$, –3, into the matrix in A –aI form, you get the following:

$$A - aI = \begin{vmatrix} 2 & -1 \\ 2 & -1 \end{vmatrix}$$

Then you have

$$\begin{vmatrix} 2 & -1 \\ 2 & -1 \end{vmatrix}\begin{vmatrix} \psi_1 \\ \psi_2 \end{vmatrix} = \begin{vmatrix} 0 \\ 0 \end{vmatrix}$$

So $2\psi_1 - \psi_2 = 0$, and $\psi_1 = \psi_2 \div 2$. And that means that, up to an arbitrary constant, the eigenvector corresponding to $a_2$ is

$$c\begin{vmatrix} 1 \\ 2 \end{vmatrix}$$

Drop the arbitrary constant:

$$\begin{vmatrix} 1 \\ 2 \end{vmatrix}$$

So the eigenvalues of this next matrix operator

$$A = \begin{vmatrix} -1 & -1 \\ 2 & -4 \end{vmatrix}$$

are $a_1 = -2$ and $a_2 = -3$. And the eigenvector corresponding to $a_1$ is

$$\begin{vmatrix} 1 \\ 1 \end{vmatrix}$$

The eigenvector corresponding to $a_2$ is

$$\begin{vmatrix} 1 \\ 2 \end{vmatrix}$$

# Preparing for the Inversion: Simplifying with Unitary Operators

Applying the inverse of an operator undoes the work the operator did:

$$A^{-1}A = AA^{-1} = I$$

Sometimes, finding the inverse of an operator is helpful, such as when you want to solve equations like $Ax = y$. Solving for $x$ is easy if you can find the inverse of A: $x = A^{-1}y$.

However, finding the inverse of a large matrix often isn't easy, so quantum physics calculations are sometimes limited to working with unitary operators, U, where the operator's inverse is equal to its adjoint, $U^{-1} = U^\dagger$. (To find the adjoint of an operator, A, you find the transpose by interchanging the rows and columns, $A^T$. Then take the complex conjugate, $A^{T*} = A^\dagger$.) This gives you the following equation:

$$U^\dagger U = UU^\dagger = I$$

The product of two unitary operators, U and V, is also unitary because

$$(UV)(VU)^\dagger = (UV)(V^\dagger U^\dagger) = U(VV^\dagger)U^\dagger = UU^\dagger = I$$

When you use unitary operators, kets and bras transform this way:

✔ $|\psi'> = U|\psi>$

✔ $|\psi'> = <\psi|U^\dagger$

And you can transform other operators using unitary operators like this:

$$A' = UAU^\dagger$$

Note that the preceding equations also mean the following:

- $|\psi> = U^\dagger|\psi'>$

- $<\psi| = <\psi'|U$

- $A = U^\dagger A'U$

Here are some properties of unitary transformations:

- If an operator is Hermitian, then its unitary transformed version, $A' = UAU^\dagger$, is also Hermitian.

- The eigenvalues of A and its unitary transformed version, $A' = UAU^\dagger$, are the same.

- Commutators that are equal to complex numbers are unchanged by unitary transformations: $[A', B'] = [A, B]$.

# Comparing Matrix and Continuous Representations

Werner Heisenberg developed the matrix-oriented view of quantum physics that you've been using so far in this chapter. It's sometimes called *matrix mechanics.* The matrix representation is fine for many problems, but sometimes you have to go past it, as you're about to see.

One of the central problems of quantum mechanics is to calculate the energy levels of a system. The energy operator is called the *Hamilitonian,* H, and finding the energy levels of a system breaks down to finding the eigenvalues of the problem:

$$H|\psi> = E|\psi>$$

Here, E is an eigenvalue of the H operator.

Here's the same equation in matrix terms:

$$\det \begin{vmatrix} H_{11}-E & H_{12} & H_{13} & H_{14} & \ldots \\ H_{21} & H_{22}-E & H_{23} & H_{24} & \ldots \\ H_{31} & H_{32} & H_{33}-E & H_{34} & \ldots \\ H_{41} & H_{42} & H_{43} & H_{44}-E & \ldots \\ & & \vdots & & \end{vmatrix} = 0$$

The allowable energy levels of the physical system are the eigenvalues E.

That's fine if you have a discrete basis of eigenvectors — if the number of energy states is finite. But what if the number of energy states is infinite? In that case, you can no longer use a discrete basis for your operators and bras and kets — you use a *continuous* basis.

## Going continuous with calculus

Representing quantum mechanics in a continuous basis is an invention of the physicist Erwin Schrödinger. In the continuous basis, summations become integrals. For example, take the following relation, where I is the identity matrix:

$$\sum_{n=1}^{\infty} |\phi_n\rangle\langle\phi_n| = I$$

It becomes the following:

$$\int_{-\infty}^{\infty} dn |\phi_n\rangle\langle\phi_n| = I$$

And every ket $|\psi\rangle$ can be expanded in a basis of other kets, $|\phi_n\rangle$, like this:

$$|\psi\rangle = \int_{-\infty}^{\infty} dn |\phi_n\rangle\langle\phi_n|\psi\rangle$$

## Doing the wave

Take a look at the position operator, R, in a continuous basis. Applying this operator gives you *r*, the position vector:

$$R|\psi\rangle = r|\psi\rangle$$

In this equation, applying the position operator to a state vector returns the locations, $r$, that a particle may be found at. You can expand any ket in the position basis like this:

$$|\psi> = \int_{-\infty}^{\infty} d^3r |r><r|\psi>$$

And this becomes

$$|\psi> = \int_{-\infty}^{\infty} d^3r \psi(r)|r><r|\psi>$$

Here's a very important thing to understand: $\psi(r) = <r|\psi>$ is the *wave function* for the state vector $|\psi>$ — it's the ket's representation in the position basis. Or in common terms, it's just a function where the quantity $|\psi(r)|^2 d^3r$ represents the probability that the particle will be found in the region $d^3r$ at $r$.

The wave function is the foundation of what's called *wave mechanics,* as opposed to matrix mechanics. What's important to realize is that when you talk about representing physical systems in wave mechanics, you don't use the basis-less bras and kets of matrix mechanics; rather, you usually use the wave function — that is, bras and kets in the position basis.

Therefore, you go from talking about $|\psi>$ to $<r|\psi>$, which equals $\psi(r)$. This wave function appears a lot in the coming chapters, and it's just a ket in the position basis. So in wave mechanics, $H|\psi> = E|\psi>$ becomes the following:

$$<r|H|\psi> = E<r|\psi>$$

You can write this as the following:

$$<r|H|\psi> = E\psi(r)$$

But what is $<r|H|\psi>$? It's equal to $H\psi(r)$. The Hamiltonian operator, H, is the total energy of the system, kinetic ($p^2/2m$) plus potential ($V(r)$) so you get the following equation:

$$H = \frac{p^2}{2m} + V(r)$$

But the momentum operator is

$$P\big|\psi> = -i\hbar\frac{\partial}{\partial x}\big|\psi>i + -i\hbar\frac{\partial}{\partial y}\big|\psi>j + -i\hbar\frac{\partial}{\partial z}\big|\psi>k$$

Therefore, substituting the momentum operator for $p$ gives you this:

$$H = \frac{-\hbar^2}{2m}\left(\frac{\partial^2}{\partial x^2} + \frac{\partial^2}{\partial y^2} + \frac{\partial^2}{\partial z^2}\right) + V(r)$$

Using the Laplacian operator, you get this equation:

$$\Delta\big|\psi> = \frac{\partial^2}{\partial x^2}\big|\psi> + \frac{\partial^2}{\partial y^2}\big|\psi> + \frac{\partial^2}{\partial z^2}\big|\psi>$$

You can rewrite this equation as the following (called the *Schrödinger equation*):

$$H\psi(r) = \frac{-\hbar^2}{2m}\Delta\psi(r) + V(r)\psi(r) = E\psi(r)$$

So in the wave mechanics view of quantum physics, you're now working with a differential equation instead of multiple matrices of elements. This all came from working in the position basis, $\psi(r)$ = <r|ψ> instead of just |ψ>.

The quantum physics in the rest of the book is largely about solving this differential equation for a variety of potentials, $V(r)$. That is, your focus is on finding the wave function that satisfies the Schrödinger equation for various physical systems. When you solve the Schrödinger equation for $\psi(r)$, you can find the allowed energy states for a physical system, as well as the probability that the system will be in a certain position state.

Note that, besides wave functions in the position basis, you can also give a wave function in the momentum basis, $\psi(p)$, or in any number of other bases.

The Heisenberg technique of matrix mechanics is one way of working with quantum physics, and it's best used for physical systems with well-defined energy states, such as harmonic oscillators. The Schrödinger way of looking at things, wave mechanics, uses wave functions, mostly in the position basis, to reduce questions in quantum physics to a differential equation.

# Part II
# Bound and Undetermined: Handling Particles in Bound States

The 5th Wave     By Rich Tennant

"Along with 'Antimatter,' and 'Dark Matter,' we've recently discovered the existence of 'Doesn't Matter,' which appears to have no effect on the universe whatsoever."

# In this part . . .

This part is where you get the lowdown on one of quantum physics' favorite topics: solving the energy levels and wave functions for particles trapped in various bound states. For example, you may have a particle trapped in a square well, which is much like having a pea in a box. Or you may have a particle in harmonic oscillation. Quantum physics is expert at handling those kinds of situations.

# Chapter 3

# Getting Stuck in Energy Wells

*W*hat's that, Lassie? Stuck in an energy well? Go get help! In this chapter, you get to see quantum physics at work, solving problems in one dimension. You see particles trapped in potential wells and solve for the allowable energy states using quantum physics. That goes against the grain in classical physics, which doesn't restrict trapped particles to any particular energy spectrum. But as you know, when the world gets microscopic, quantum physics takes over.

The equation of the moment is the Schrödinger equation (derived in Chapter 2), which lets you solve for the wave function, $\psi(x)$, and the energy levels, E:

$$\frac{-\hbar^2}{2m}\Delta \psi(r) + V(r)\psi(r) = E\psi(r)$$

## Looking into a Square Well

A *square well* is a potential (that is, a potential energy well) that forms a square shape, as you can see in Figure 3-1.

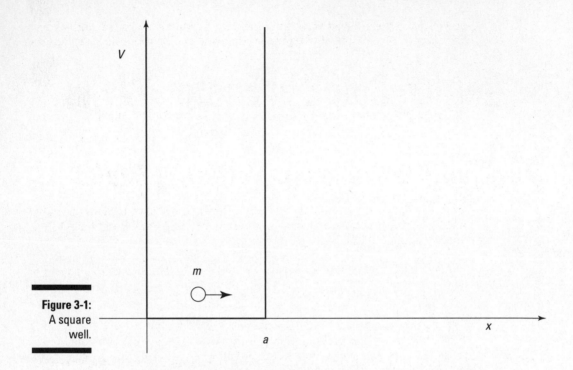

**Figure 3-1:**
A square
well.

The potential, or V($x$), goes to infinity at $x = 0$ and $x > a$ (where $x$ is distance), like this:

- ✔ V($x$) = ∞, where $x < 0$
- ✔ V($x$) = 0, where $0 \leq x \leq a$
- ✔ V($x$) = ∞, where $x > a$

Using square wells, you can trap particles. If you put a particle into a square well with a limited amount of energy, it'll be trapped because it can't overcome the infinite potential at either side of the square well. Therefore, the particle has to move inside the square well.

So does the particle just sort of roll around on the bottom of the square well? Not exactly. The particle is in a bound state, and its wave function depends on its energy. The wave function isn't complicated:

$$\psi(x) = \left(\sqrt{2}\Big/a\right) \sin \frac{n\pi}{a} x \qquad n = 1, 2, 3 \ldots$$

So you have the allowed wave functions for the states $n$ = 1, 2, 3, and so on. The energy of the allowable bound states are given by the following equation:

$$E = \frac{\hbar^2 \pi^2}{2ma^2} n^2 \qquad n = 1, 2, 3 \ldots$$

The rest of this chapter shows you how to solve problems like this one.

# Trapping Particles in Potential Wells

Take a look at the potential in Figure 3-2. Notice the dip, or *well*, in the potential, which means that particles can be trapped in it if they don't have too much energy.

The particle's kinetic energy summed with its potential energy is a constant, equal to its total energy:

$$\frac{p^2}{2m} + V = E$$

If its total energy is less than $V_1$, the particle will be trapped in the potential well, you see in Figure 3-2; to get out of the well, the particle's kinetic energy would have to become negative to satisfy the equation, which is impossible.

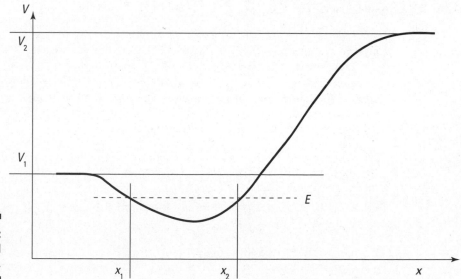

**Figure 3-2:**
A potential
well.

In this section, you take a look at the various possible states that a particle with energy E can take in the potential given by Figure 3-2. Quantum-mechanically speaking, those states are of two kinds — bound and unbound. This section looks at them in overview.

## Binding particles in potential wells

*Bound states* happen when the particle isn't free to travel to infinity — it's as simple as that. In other words, the particle is confined to the potential well.

A particle traveling in the potential well you see in Figure 3-2 is bound if its energy, E, is less than both $V_1$ and $V_2$. In that case, the particle moves between $x_1$ and $x_2$. A particle trapped in such a well is represented by a wave function, and you can solve the Schrödinger equation for the allowed wave functions and the allowed energy states. You need to use two boundary conditions (the Schrödinger equation is a second-order differential equation) to solve the problem completely.

Bound states are *discrete* — that is, they form an energy spectrum of discrete energy levels. The Schrödinger equation gives you those states. In addition, in one-dimensional problems, the energy levels of a bound state are not degenerate — that is, no two energy levels are the same in the entire energy spectrum.

## Escaping from potential wells

If a particle's energy, E, is greater than the potential $V_1$ in Figure 3-2, the particle can escape from the potential well. There are two possible cases: $V_1 < E < V_2$ and $E > V_2$. This section looks at them separately.

### Case 1: Energy between the two potentials ($V_1 < E < V_2$)

If $V_1 < E < V_2$, the particle in the potential well has enough energy to overcome the barrier on the left but not on the right. The particle is thus free to move to negative infinity, so its allowed $x$ region is between $-\infty$ and $x_1$.

Here, the allowed energy values are continuous, not discrete, because the particle isn't completely bound. The energy eigenvalues are not degenerate — that is, no two energy eigenvalues are the same (see Chapter 2 for more on eigenvalues).

The Schrödinger equation is a second-order differential equation, so it has two linearly independent solutions; however, in this case, only one of those solutions is physical and doesn't diverge.

The wave equation in this case turns out to oscillate for $x < x_2$ and to decay rapidly for $x > x_2$.

### Case 2: Energy greater than the higher potential (E > V₂)

If $E > V_2$, the particle isn't bound at all and is free to travel from negative infinity to positive infinity.

The energy spectrum is continuous and the wave function turns out to be a sum of a function moving to the right and one moving to the left. The energy levels of the allowed spectrum are therefore doubly degenerate.

That's all the overview you need — time to start solving the Schrödinger equation for various different potentials, starting with the easiest of all: infinite square wells.

# Trapping Particles in Infinite Square Potential Wells

Infinite square wells, in which the walls go to infinity, are a favorite in physics problems. You explore the quantum physics take on these problems in this section.

## Finding a wave-function equation

Take a look at the infinite square well that appears back in Figure 3-1. Here's what that square well looks like:

- $V(x) = \infty$, where $x < 0$
- $V(x) = 0$, where $0 \le x \le a$
- $V(x) = \infty$, where $x > a$

The Schrödinger equation looks like this in three dimensions:

$$\frac{-\hbar^2}{2m}\Delta\psi(r)+V(r)\psi(r)=E\psi(r)$$

Writing out the Schrödinger equation gives you the following:

$$\frac{-\hbar^2}{2m}\left(\frac{\partial^2}{\partial x^2}+\frac{\partial^2}{\partial y^2}+\frac{\partial^2}{\partial z^2}\right)\psi(r)+V(r)\psi(r)=E\psi(r)$$

You're interested in only one dimension — $x$ (distance) — in this chapter, so the Schrödinger equation looks like

$$\frac{-\hbar^2}{2m}\frac{d^2}{dx^2}\psi(x) + V(x)\psi(x) = E\psi(x)$$

Because V($x$) = 0 inside the well, the equation becomes

$$\frac{-\hbar^2}{2m}\frac{d^2}{dx^2}\psi(x) = E\psi(x)$$

And in problems of this sort, the equation is usually written as

$$\frac{d^2}{dx^2}\psi(x) + k^2\psi(x) = 0$$

where $k^2 = \dfrac{2mE}{\hbar^2}$ ($k$ is the wave number).

So now you have a second-order differential equation to solve for the wave function of a particle trapped in an infinite square well.

You get two independent solutions because this equation is a second-order differential equation:

$\psi_1(x)$ = A sin($kx$)

$\psi_2(x)$ = B cos($kx$)

A and B are constants that are yet to be determined.

The general solution of $\dfrac{d^2}{dx^2}\psi(x) + k^2\psi(x) = 0$ is the sum of $\psi_1(x)$ and $\psi_2(x)$:

$\psi(x)$ = A sin($kx$) + B cos($kx$)

## Determining the energy levels

The equation $\psi(x)$ = A sin($kx$) + B cos($kx$) tells you that you have to use the boundary conditions to find the constants A and B (the preceding section explains how to derive the equation). What are the boundary conditions? The wave function must disappear at the boundaries of an infinite square well, so

✔ $\psi(0) = 0$

✔ $\psi(a) = 0$

The fact that $\psi(0) = 0$ tells you right away that B must be zero because $\cos(0) = 1$. And the fact that $\psi(a) = 0$ tells you that $\psi(a) = A \sin(ka) = 0$. Because sine is zero when its argument is a multiple of $\pi$, this means that

$$ka = n\pi \qquad n = 1, 2, 3 \dots$$

Note that although $n = 0$ is technically a solution, it yields $\psi(0) = 0$, so it's not a physical solution — the physical solutions begin with $n = 1$.

This equation can also be written as

$$k = \frac{n\pi}{a} \qquad n = 1, 2, 3 \dots$$

And because $k^2 = 2mE/\hbar^2$, you have the following equation, where $n = 1, 2, 3, \dots$ — those are the allowed energy states. These are quantized states, corresponding to the quantum numbers 1, 2, 3, and so on:

$$\frac{2mE}{\hbar^2} = \frac{n^2\pi^2}{a^2}$$

$$E = \frac{n^2\hbar^2\pi^2}{2ma^2}$$

Note that the first physical state corresponds to $n = 1$, which gives you this next equation:

$$E = \frac{\hbar^2\pi^2}{2ma^2}$$

This is the lowest physical state that the particles can occupy. Just for kicks, put some numbers into this, assuming that you have an electron, mass $9.11 \times 10^{-31}$ kilograms, confined to an infinite square well of width of the order of *Bohr radius* (the average radius of an electron's orbit in a hydrogen atom), about $10^{-10}$ meters.

$E = \frac{\hbar^2\pi^2}{2ma^2}$ gives you this energy for the ground state:

$$\frac{\left(1.05 \times 10^{-34}\right)^2 \left(3.14\right)^2}{2\left(9.11 \times 10^{-31}\right)\left(10^{-10}\right)^2} = 6.00 \times 10^{-19} \text{ Joules}$$

That's a very small amount, about 4.0 electron volts (eV — the amount of energy one electron gains falling through 1 volt). Even so, it's already on the order of the energy of the ground state of an electron in the ground state of a hydrogen atom (13.6 eV), so you can say you're certainly in the right quantum physics ballpark now.

## Normalizing the wave function

Okay, you have this for the wave equation for a particle in an infinite square well:

$$\psi(a) = A \sin\left(\frac{n\pi x}{a}\right)$$

The wave function is a sine wave, going to zero at $x = 0$ and $x = a$. You can see the first two wave functions plotted in Figure 3-3.

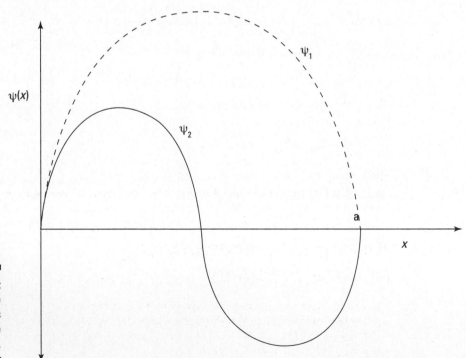

**Figure 3-3:**
Wave functions in a square well.

Normalizing the wave function lets you solve for the unknown constant A. In a *normalized* function, the probability of finding the particle between $x$ and $dx$, $|\psi(x)|^2dx$, adds up to 1 when you integrate over the whole square well, $x = 0$ to $x = a$:

$$1 = \int_0^a |\psi(x)|^2 dx$$

Substituting for $\psi(x)$ gives you the following:

$$1 = |A|^2 \int_0^a \sin^2\left(\frac{n\pi x}{a}\right)dx$$

Here's what the integral in this equation equals:

$$\int_0^a \sin^2\left(\frac{n\pi x}{a}\right)dx = \frac{a}{2}$$

So from the previous equation, $1 = |A|^2\left(\frac{a}{2}\right)$. Solve for A:

$$A = \left(\frac{2}{a}\right)^{1/2}$$

Therefore, here's the normalized wave equation with the value of A plugged in:

$$\psi(x) = \left(\frac{2}{a}\right)^{1/2} \sin\left(\frac{n\pi x}{a}\right) \qquad n = 1, 2, 3 \dots$$

And that's the normalized wave function for a particle in an infinite square well.

## Adding time dependence to wave functions

Now how about seeing how the wave function for a particle in an infinite square well evolves with time? The Schrödinger equation looks like this:

$$\frac{-\hbar^2}{2m}\Delta\psi(r) + V(r)\psi(r) = E\psi(r)$$

You can also write the Schrödinger equation this way, where H is the Hermitian Hamiltonian operator:

$$H\psi(r) = E\psi(r)$$

That's actually the *time-independent* Schrödinger equation. The *time-dependent* Schrödinger equation looks like this:

$$i\hbar\frac{\partial}{\partial t}\psi(r,\,t)=H\psi(r,\,t)$$

Combining the preceding three equations gives you the following, which is another form of the time-dependent Schrödinger equation:

$$i\hbar\frac{\partial}{\partial t}\psi(r,\,t)=\frac{-\hbar^2}{2m}\Delta\psi(r,\,t)+V(r,\,t)\psi(r,\,t)$$

And because you're dealing with only one dimension, *x,* this equation becomes

$$i\hbar\frac{\partial}{\partial t}\psi(x,\,t)=\frac{-\hbar^2}{2m}\frac{d}{dx}\psi(x,\,t)+V(x,\,t)\psi(x,\,t)$$

This is simpler than it looks, however, because the potential doesn't change with time. In fact, because E is constant, you can rewrite the equation as

$$i\hbar\frac{\partial}{\partial t}\psi(x,\,t)=E\psi(x,\,t)$$

That equation makes life a lot simpler — it's easy to solve the time-dependent Schrödinger equation if you're dealing with a constant potential. In this case, the solution is

$$\psi(x,\,t)=\psi(x)e^{-iEt/\hbar}$$

Neat. When the potential doesn't vary with time, the solution to the time-dependent Schrödinger equation simply becomes $\psi(x)$, the spatial part, multiplied by $e^{-iEt/\hbar}$, the time-dependent part.

So when you add in the time-dependent part to the time-independent wave function, you get the time-dependent wave function, which looks like this:

$$\psi(x,\, t) = \left(\frac{2}{a}\right)^{\frac{1}{2}} \sin\left(\frac{n\pi x}{a}\right) e^{-iEt/\hbar} \quad n = 1,\, 2,\, 3\, \dots$$

The energy of the $n$th quantum state is

$$E = \frac{n^2 \hbar^2 \pi^2}{2ma^2} \quad n = 1,\, 2,\, 3\, \dots$$

Therefore, the result is

$$\psi(x,\, t) = \left(\frac{2}{a}\right)^{\frac{1}{2}} \sin\left(\frac{n\pi x}{a}\right) \exp\left(\frac{-in^2 \hbar^2 \pi^2 t}{2ma^2}\right) \quad n = 1,\, 2,\, 3\, \dots$$
where $\exp(x) = e^x$.

# Shifting to symmetric square well potentials

The standard infinite square well looks like this:

- $V(x) = \infty$, where $x < 0$
- $V(x) = 0$, where $0 \leq x \leq a$
- $V(x) = \infty$, where $x > a$

But what if you want to shift things so that the square well is symmetric around the origin instead? That is, you move the square well so that it extends from $-a/2$ to $a/2$? Here's what the new infinite square well looks like in this case:

- $V(x) = \infty$, where $x < -a/2$
- $V(x) = 0$, where $-a/2 \leq x \leq a/2$
- $V(x) = \infty$, where $x > a/2$

You can translate from this new square well to the old one by adding $a/2$ to $x$, which means that you can write the wave function for the new square well in this equation like the following:

$$\psi(x) = \left(\frac{2}{a}\right)^{\frac{1}{2}} \sin\left(\frac{n\pi}{a}\left(x+\frac{a}{2}\right)\right) \quad n = 1, 2, 3 \ldots$$

Doing a little trig gives you the following equations:

$$\psi(x) = \left(\frac{2}{a}\right)^{\frac{1}{2}} \cos\left(\frac{n\pi(x)}{a}\right) \quad n = 1, 3, 5 \ldots$$

$$\psi(x) = \left(\frac{2}{a}\right)^{\frac{1}{2}} \sin\left(\frac{n\pi(x)}{a}\right) \quad n = 2, 4, 6 \ldots$$

So as you can see, the result is a mix of sines and cosines. The bound states are these, in increasing quantum order:

✔ $\psi_1(x) = \left(\frac{2}{a}\right)^{\frac{1}{2}} \cos\left(\frac{\pi x}{a}\right)$

✔ $\psi_2(x) = \left(\frac{2}{a}\right)^{\frac{1}{2}} \sin\left(\frac{2\pi x}{a}\right)$

✔ $\psi_2(x) = \left(\frac{2}{a}\right)^{\frac{1}{2}} \cos\left(\frac{2\pi x}{a}\right)$

✔ $\psi_4(x) = \left(\frac{2}{a}\right)^{\frac{1}{2}} \sin\left(\frac{4\pi x}{a}\right)$

And so on.

Note that the cosines are symmetric around the origin: $\psi(x) = \psi(-x)$. The sines are anti-symmetric: $-\psi(x) = \psi(-x)$.

# Limited Potential: Taking a Look at Particles and Potential Steps

Truly infinite potentials (which I discuss in the previous sections) are hard to come by. In this section, you look at some real-world examples, where the potential is set to some finite $V_0$, not infinity. For example, take a look at the situation in Figure 3-4. There, a particle is traveling toward a potential step. Currently, the particle is in a region where $V = 0$, but it'll soon be in the region $V = V_0$.

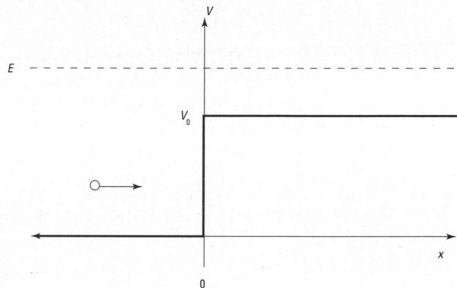

**Figure 3-4:**
A potential
step, $E > V_0$.

There are two cases to look at here in terms of E, the energy of the particle:

- **E > V$_0$:** Classically, when $E > V_0$, you expect the particle to be able to continue on to the region $x > 0$.

- **E < V$_0$:** When $E < V_0$, you'd expect the particle to bounce back and not be able to get to the region $x > 0$ at all.

In this section, you start by taking a look at the case where the particle's energy, E, is greater than the potential $V_0$, as shown in Figure 3-4; then you take a look at the case where $E < V_0$.

## Assuming the particle has plenty of energy

Start with the case where the particle's energy, E, is greater than the potential $V_0$. From a quantum physics point of view, here's what the Schrödinger equation would look like:

- **For the region $x < 0$:** $\dfrac{d^2\psi_1}{dx^2}(x) + k_1^2\psi_1(x) = 0$

  Here, $k_1^2 = 2m\text{E}\big/{\hbar^2}$.

✔ **For the region $x > 0$:** $\dfrac{d^2\psi_2}{dx^2}(x) + k_2{}^2\psi_2(x) = 0$

In this equation, $k_2{}^2 = \dfrac{2m(E - V_0)}{\hbar^2}$.

In other words, $k$ is going to vary by region, as you see in Figure 3-5.

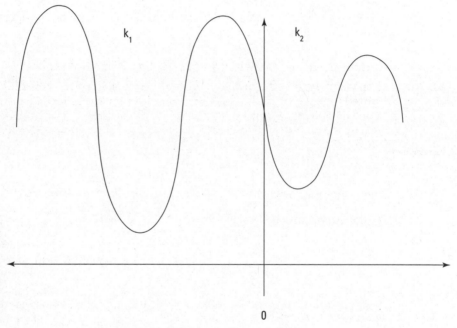

**Figure 3-5:**
The value of
$k$ by region,
where
$E > V_0$.

Treating the first equation as a second-order differential equation, you can see that the most general solution is the following:

$\psi_1(x) = Ae^{ik_1x} + Be^{-ik_1x}$, where $x < 0$

And for the region $x > 0$, solving the second equation gives you this:

$\psi_2(x) = Ce^{ik_2x} + De^{-ik_2x}$, where $x > 0$

Note that $e^{ikx}$ represents plane waves traveling in the $+x$ direction, and $e^{-ikx}$ represents plane waves traveling in the $-x$ direction.

What this solution means is that waves can hit the potential step from the left and be either transmitted or reflected. Given that way of looking at the problem, you may note that the wave can be reflected only going to the right, not to the left, so D must equal zero. That makes the wave equation become the following:

> ✔ **Where** $x < 0$: $\psi_1(x) = Ae^{ik_1x} + Be^{-ik_1x}$
>
> ✔ **Where** $x > 0$: $\psi_2(x) = Ce^{ik_2x}$

The term $Ae^{ik_1x}$ represents the incident wave, $Be^{-ik_1x}$ is the reflected wave, and $Ce^{ik_2x}$ is the transmitted wave.

### Calculating the probability of reflection or transmission

You can calculate the probability that the particle will be reflected or transmitted through the potential step by calculating the *reflection* and *transmission coefficients*. If $J_r$ is the reflected current density, $J_i$ is the incident current density, and $J_t$ is the transmitted current density, then R, the reflection coefficient is

$$R = \frac{J_r}{J_i}$$

T, the transmission coefficient, is

$$T = \frac{J_t}{J_i}$$

You now have to calculate $J_r$, $J_i$, and $J_t$. Actually, that's not so hard — start with $J_i$. Because the incident part of the wave is $\psi_i(x) = Ae^{ik_1x}$, the incident current density is

$$J_i = \frac{i\hbar}{2m}\left[\psi_i(x)\frac{d\psi_i^*(x)}{dx} - \psi_i^*(x)\frac{d\psi_i(x)}{dx}\right]$$

And this just equals $\frac{\hbar k_1}{m}\left|A\right|^2$. $J_r$ and $J_t$ work in the same way:

$$J_r = \frac{\hbar k_1}{m}\left|B\right|^2$$

$$J_t = \frac{\hbar k_2}{m}\left|C\right|^2$$

So you have this for the reflection coefficient:

$$R = \frac{J_r}{J_i} = \frac{|B|^2}{|A|^2}$$

T, the transmission coefficient, is

$$T = \frac{J_t}{J_i} = \frac{|C|^2}{|A|^2}$$

## Finding A, B, and C

So how do you figure out the constants A, B, and C? You do that as you figure out the coefficients with the infinite square well potential — with boundary conditions (see the earlier section "Trapping Particles in Infinite Square Well Potentials"). However, here, you can't necessarily say that $\psi(x)$ goes to zero, because the potential is no longer infinite. Instead, the boundary conditions are that $\psi(x)$ and $d\psi(x)/dx$ are continuous across the potential step's boundary. In other words,

  ✔ $\psi_1(0) = \psi_2(0)$

  ✔ $\frac{d\psi_1}{dx}(0) = \frac{d\psi_2}{dx}(0)$

You know the following:

  ✔ **Where $x < 0$:** $\psi_1(x) = Ae^{ik_1x} + Be^{-ik_1x}$

  ✔ **Where $x > 0$:** $\psi_2(x) = Ce^{ik_2x}$

Therefore, plugging these two equations into $\psi_1(0) = \psi_2(0)$ gives you A + B = C. And plugging them into $\frac{d\psi_1}{dx}(0) = \frac{d\psi_2}{dx}(0)$ gives you

$$k_1A - k_1B = k_2C$$

Solving for B in terms of A gives you this result:

$$B = \frac{k_1 - k_2}{k_1 + k_2} A$$

Solving for C in terms of A gives you

$$C = \frac{2k_1}{k_1 + k_2} A$$

You can then calculate A from the normalization condition of the wave function:

$$1 = \int |\psi(x)|^2 \, dx$$

But you don't actually need A, because it drops out of the ratios for the reflection and transmission coefficients, R and T. In particular,

$$R = \frac{|B|^2}{|A|^2}$$

$$T = \frac{|C|^2}{|A|^2}$$

Therefore,

$$R = \frac{(k_1 - k_2)^2}{(k_1 + k_2)^2}$$

$$T = \frac{4k_1 k_2}{(k_1 + k_2)^2}$$

That's an interesting result, and it disagrees with classical physics, which says that there should be no particle reflection at all. As you can see, if $k_1 \neq k_2$, then there will indeed be particle reflection.

Note that as $k_1$ goes to $k_2$, R goes to 0 and T goes to 1, which is what you'd expect.

So already you have a result that differs from the classical — the particle can be reflected at the potential step. That's the wave-like behavior of the particle coming into play again.

# Assuming the particle doesn't have enough energy

Okay, now try the case where $E < V_0$ when there's a potential step, as shown in Figure 3-6. In this case, the particle doesn't have enough energy to make it into the region $x > 0$, according to classical physics. See what quantum physics has to say about it.

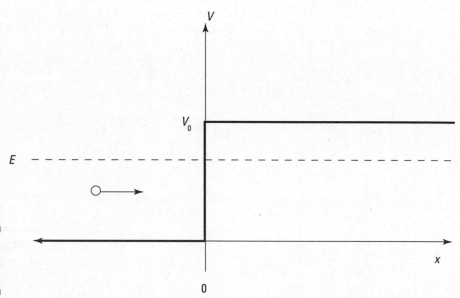

**Figure 3-6:**
A potential
step, $E < V_0$.

Tackle the region $x < 0$ first. There, the Schrödinger equation would look like this:

$$\frac{d^2\psi_1}{dx^2}(x) + k_1^2 \psi_1(x) = 0$$

where $k_1^2 = \dfrac{2mE}{\hbar^2}$ .

You know the solution to this from the previous discussion on potential steps (see "Limited Potential: Taking a Look at Particles and Potential Steps"):

$$\psi_1(x) = Ae^{ik_1x} + Be^{-ik_1x} \qquad x < 0$$

Okay, but what about the region $x > 0$? That's a different story. Here's the Schrödinger equation:

$$\frac{d^2\psi_2}{dx^2}(x) - k^2\psi_2(x) = 0 \quad \text{(where } x > 0\text{)}$$

where $k^2 = \dfrac{2m(E - V_0)}{\hbar^2}.$

But hang on; $E - V_0$ is less than zero, which would make $k$ imaginary, which is impossible physically. So change the sign in the Schrödinger equation from plus to minus:

$$\frac{d^2\psi_2}{dx^2}(x) - k^2\psi_2(x) = 0 \quad x > 0$$

And use the following for $k_2$ (note that this is positive if $E < V_0$):

$$k_2^2 = \frac{2m(V_0 - E)}{\hbar^2}$$

Okay, so now you have to solve the differential

$\dfrac{d^2\psi_2}{dx^2}(x) - k^2\psi_2(x) = 0$     (where $x > 0$). There are two linearly independent solutions:

- ✔ $\psi(x) = Ce^{-k_2 x}$
- ✔ $\psi(x) = De^{k_2 x}$

And the general solution to $\dfrac{d^2\psi_2}{dx^2}(x) - k^2\psi_2(x) = 0$     (where $x > 0$) is

$$\psi_2(x) = Ce^{-k_2 x} + De^{-k_2 x} \quad x > 0$$

However, wave functions must be finite everywhere, and the second term is clearly not finite as $x$ goes to infinity, so D must equal zero (note that if $x$ goes to negative infinity, the first term also diverges, but because the potential step is limited to $x > 0$, that isn't a problem). Therefore, here's the solution for $x > 0$:

$$\psi_2(x) = Ce^{-k_2 x} \quad x > 0$$

So your wave functions for the two regions are

$$\psi_1(x) = Ae^{ik_1x} + Be^{-ik_1x} \quad x < 0$$

$$\psi_2(x) = Ce^{-k_2x} \quad\quad x > 0$$

Putting this in terms of the incident, reflected, and transmitted wave functions, $\psi_i(x)$, $\psi_r(x)$, and $\psi_t(x)$, you have the following:

- ✔ $\psi_i(x) = Ae^{ik_1x}$
- ✔ $\psi_r(x) = Be^{-ik_1x}$
- ✔ $\psi_t(x) = Ce^{-k_2x}$

### Finding transmission and reflection coefficients

Now you can figure out the reflection and transmission coefficients, R and T (as you do for the case $E > V_0$ in the earlier section "Assuming the particle has plenty of energy"):

$$R = \frac{J_r}{J_i}$$

$$T = \frac{J_t}{J_i}$$

Actually, this is very easy in this case; take a look at $J_t$:

$$J_t = \frac{i\hbar}{2m}\left[\psi_t(x)\frac{d\psi_t^*(x)}{dx} - \psi_t^*(x)\frac{d\psi_t(x)}{dx}\right]$$

But because $\psi_t(x) = Ce^{-k_2x}$, $\psi_t(x)$ is completely real, which means that in this case, the following is true:

$$J_t = \frac{i\hbar}{2m}\left[\psi_t(x)\frac{d\psi_t(x)}{dx} - \psi_t(x)\frac{d\psi_t(x)}{dx}\right]$$

And this equation, of course, is equal to zero.

So $J_t = 0$; therefore, T = 0. If T = 0, then R must equal 1. That means that you have a complete reflection, just as in the classical solution.

### The nonzero solution: Finding a particle in x > 0

Despite the complete reflection, there's a difference between the mathematical and classical solution: There actually is a nonzero chance of finding the particle in the region $x > 0$. To see that, take a look at the probability density for $x > 0$, which is

$$P(x) = |\psi_t(x)|^2$$

Plugging in for the wave function $\psi_t(x)$ gives you

$$P(x) = |\psi_t(x)|^2 = |C|^2 e^{-2k_2 x}$$

You can use the continuity conditions to solve for C in terms of A:

➤ $\psi_1(0) = \psi_2(0)$

➤ $\dfrac{d\psi_1}{dx}(0) = \dfrac{d\psi_2}{dx}(0)$

Using the continuity conditions gives you the following:

$$P(x) = |C|^2 e^{-2k_2 x} = \frac{4k_1^2 |A| e^{-2k_2 x}}{k_1^2 + k_2^2}$$

This does fall quickly to zero as $x$ gets large, but near $x = 0$, it has a nonzero value.

You can see what the probability density looks like for the $E < V_0$ case of a potential step in Figure 3-7.

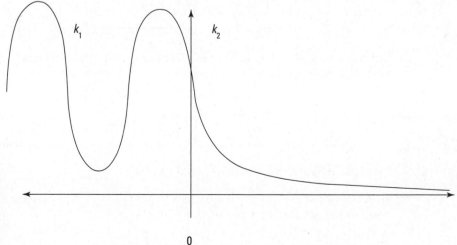

**Figure 3-7:**
The value of
k by region,
$E < V_0$.

$k_1$

$k_2$

0

Okay, you've taken care of infinite square wells and potential steps. Now what about the case where the potential step doesn't extend out to infinity but is itself bounded? That brings you to *potential barriers,* which I discuss in the next section.

# Hitting the Wall: Particles and Potential Barriers

What if the particle could work its way through a potential step — that is, the step was of limited extent? Then you'd have a potential barrier, which is set up something like this:

- ✔ $V(x) = 0$, where $x < 0$
- ✔ $V(x) = V_0$, where $0 \le x \le a$
- ✔ $V(x) = 0$, where $x > a$

You can see what this potential looks like in Figure 3-8.

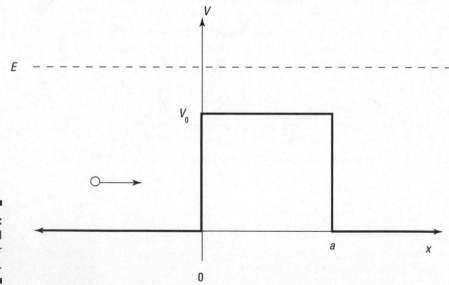

**Figure 3-8:**
A potential
barrier
$E > V_0$.

In solving the Schrödinger equation for a potential barrier, you have to consider two cases, corresponding to whether the particle has more or less energy than the potential barrier. In other words, if E is the energy of the incident particle, the two cases to consider are $E > V_0$ and $E < V_0$. This section starts with $E > V_0$.

## Getting through potential barriers when $E > V_0$

In the case where $E > V_0$, the particle has enough energy to pass through the potential barrier and end up in the $x > a$ region. This is what the Schrödinger equation looks like:

- **For the region $x < 0$:** $\dfrac{d^2\psi_1}{dx^2}(x) + k_1{}^2\psi_1(x) = 0$

  where $k_1{}^2 = \dfrac{2m\mathrm{E}}{\hbar^2}$

- **For the region $0 \leq x \leq a$:** $\dfrac{d^2\psi_2}{dx^2}(x) + k_2{}^2\psi_2(x) = 0$

  where $k_2{}^2 = \dfrac{2m(\mathrm{E} - \mathrm{V}_0)}{\hbar^2}$

- **For the region $x > a$:** $\dfrac{d^2\psi_3}{dx^2}(x) + k_1{}^2\psi_3(x) = 0$

  where $k_1{}^2 = \dfrac{2m\mathrm{E}}{\hbar^2}$

The solutions for $\psi_1(x)$, $\psi_2(x)$, and $\psi_3(x)$ are the following:

- **Where $x < 0$:** $\psi_1(x) = Ae^{ik_1 x} + Be^{-ik_1 x}$
- **Where $0 \leq x \leq a$:** $\psi_2(x) = Ce^{ik_2 x} + De^{-ik_2 x}$
- **Where $x > a$:** $\psi_3(x) = Ee^{ik_1 x} + Fe^{-ik_1 x}$

  In fact, because there's no leftward traveling wave in the $x > a$ region, $F = 0$, so $\psi_3(x) = Ee^{ik_1 x}$.

So how do you determine A, B, C, D, and E? You use the continuity conditions, which work out here to be the following:

$$\psi_1(0) = \psi_2(0)$$

$$\frac{d\psi_1}{dx}(0) = \frac{d\psi_2}{dx}(0)$$

$$\psi_2(0) = \psi_3(0)$$

$$\frac{d\psi_2}{dx}(0) = \frac{d\psi_3}{dx}(0)$$

Okay, from these equations, you get the following:

- A + B = C + D
- $ik_1(A - B) = ik_2(C - D)$
- $Ce^{ik_2a} + De^{-ik_2a} = Ee^{ik_1a}$
- $ik_2Ce^{ik_2a} - ik_2De^{-ik_2a} = ik_1Ee^{ik_1a}$

So putting all of these equations together, you get this for the coefficient E in terms of A:

$$E = 4k_1k_2Ae^{-ik_1a}\left[4k_1k_2\cos(k_2a) - 2i\left(k_1^2 + k_2^2\right)\sin(k_2a)\right]^{-1}$$

Wow. So what's the transmission coefficient, T? Well, T is

$$T = \frac{|E|^2}{|A|^2}$$

And this works out to be

$$T = \left[1 + \frac{1}{4}\left(\frac{k_1^2 - k_2^2}{k_1k_2}\right)^2 \sin^2(k_2a)\right]^{-1}$$

Whew! Note that as $k_1$ goes to $k_2$, T goes to 1, which is what you'd expect.

So how about R, the reflection coefficient? I'll spare you the algebra; here's what R equals:

$$R = \left[ 1 + \frac{4\left(\dfrac{E}{V_0}\right)\left(\dfrac{E}{V_0}-1\right)}{\sin^2\left(a\left(\dfrac{2mV_0}{\hbar^2}\right)^{1/2}\right)\left(\dfrac{E}{V_0}-1\right)^{1/2}} \right]$$

You can see what the $E > V_0$ probability density, $|\psi(x)|^2$, looks like for the potential barrier in Figure 3-9.

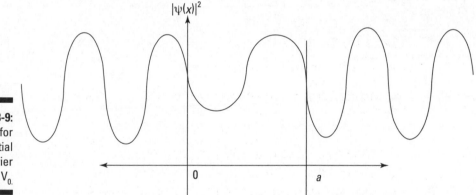

**Figure 3-9:**
$|\psi(x)|^2$ for a potential barrier $E > V_0$.

That completes the potential barrier when $E > V_0$.

## Getting through potential barriers, even when $E < V_0$

What happens if the particle doesn't have as much energy as the potential of the barrier? In other words, you're now facing the situation you see in Figure 3-10.

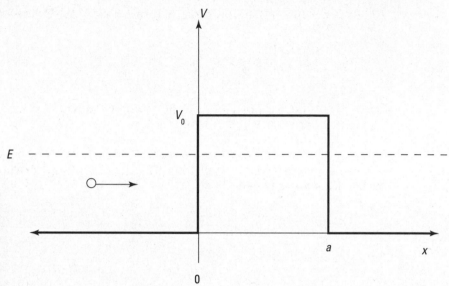

**Figure 3-10:**
A potential
barrier
$E < V_0$.

Now the Schrödinger equation looks like this:

- **For the region $x < 0$:** $\psi_1(x) = Ae^{ik_1x} + Be^{-ik_1x}$

- **For the region $0 \le x \le a$:** $\dfrac{d^2\psi_2}{dx^2}(x) + k^2\psi_2(x) = 0$

  where $k^2 = \dfrac{2m(E - V_0)}{\hbar^2}$.

But now $E - V_0$ is less than 0, which would make $k$ imaginary. And that's impossible physically. So change the sign in the Schrödinger equation from plus to minus:

$$\frac{d^2\psi_2}{dx^2}(x) - k_2{}^2\psi_2(x) = 0$$

And use this for $k_2$: $k_2{}^2 = \dfrac{2m(E - V_0)}{\hbar^2}$.

- **For the region $x > a$:** $\dfrac{d^2\psi_3}{dx^2}(x) - k_1{}^2\psi_3(x) = 0$

  where $k_1{}^2 = \dfrac{2mE}{\hbar^2}$.

All this means that the solutions for $\psi_1(x)$, $\psi_2(x)$, and $\psi_3(x)$ are the following:

- **Where $x < 0$:** $\psi_1(x) = Ae^{ik_1x} + Be^{-ik_1x}$
- **Where $0 \leq x \leq a$:** $\psi_2(x) = Ce^{k_2x} + De^{-k_2x}$
- **Where $x > a$:** $\psi_3(x) = Ee^{ik_1x} + Fe^{-ik_1x}$

    In fact, there's no leftward traveling wave in the region $x > a$; F = 0, so $\psi_3(x)$ is $\psi_3(x) = Ee^{ik_1x}$.

This situation is similar to the case where E > $V_0$, except for the region $0 \leq x \leq a$. The wave function oscillates in the regions where it has positive energy, $x < 0$ and $x > a$, but is a decaying exponential in the region $0 \leq x \leq a$.

You can see what the probability density, $|\psi(x)|^2$, looks like in Figure 3-11.

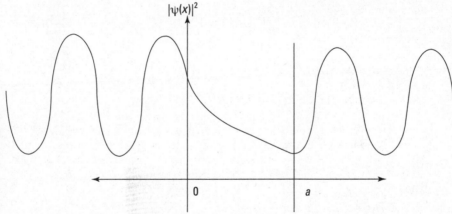

**Figure 3-11:** $|\psi(x)|^2$ for a potential barrier E < $V_0$.

## Finding the reflection and transmission coefficients

How about the reflection and transmission coefficients, R and T? Here's what they equal:

$$R = \frac{|B|^2}{|A|^2}$$

$$T = \frac{|E|^2}{|A|^2}$$

As you may expect, you use the continuity conditions to determine A, B, and E:

- $\psi_1(0) = \psi_2(0)$

- $\dfrac{d\psi_1}{dx}(0) = \dfrac{d\psi_2}{dx}(0)$

- $\psi_2(a) = \psi_3(a)$

- $\dfrac{d\psi_2}{dx}(a) = \dfrac{d\psi_3}{dx}(a)$

A fair bit of algebra and trig is involved in solving for R and T; here's what R and T turn out to be:

$$R = \left(\frac{k_1^2 + k_2^2}{k_1 k_2}\right)^2 \sinh^2(k_2 a)\left[4\cosh^2(k_2 a) + \left(\frac{k_1^2 + k_2^2}{k_1 k_2}\right)^2 \sinh^2(k_2 a)\right]^{-1}$$

$$T = 4\left[4\cosh^2(k_2 a) + \left(\frac{k_1^2 + k_2^2}{k_1 k_2}\right)^2 \sinh^2(k_2 a)\right]^{-1}$$

Despite the equation's complexity, it's amazing that the expression for T can be nonzero. Classically, particles can't enter the forbidden zone $0 \le x \le a$ because $E < V_0$, where $V_0$ is the potential in that region; they just don't have enough energy to make it into that area.

## Tunneling through

Quantum mechanically, the phenomenon where particles can get through regions that they're classically forbidden to enter is called *tunneling*. Tunneling is possible because in quantum mechanics, particles show wave properties.

Tunneling is one of the most exciting results of quantum physics — it means that particles can actually get through classically forbidden regions because of the spread in their wave functions. This is, of course, a microscopic effect — don't try to walk through any closed doors — but it's a significant one. Among other effects, tunneling makes transistors and integrated circuits possible.

You can calculate the transmission coefficient, which tells you the probability that a particle gets through, given a certain incident intensity, when tunneling is involved. Doing so is relatively easy in the preceding section because the barrier that the particle has to get through is a square barrier. But in general, calculating the transmission coefficient isn't so easy. Read on.

### Getting the transmission with the WKB approximation

The way you generally calculate the transmission coefficient is to break up the potential you're working with into a succession of square barriers and to sum them. That's called the *Wentzel-Kramers-Brillouin* (WKB) approximation — treating a general potential, V(x), as a sum of square potential barriers.

The result of the WKB approximation is that the transmission coefficient for an arbitrary potential, V(x), for a particle of mass $m$ and energy E is given by this expression (that is, as long as V(x) is a smooth, slowly varying function):

$$T \sim \exp\left[ -\left( \frac{2}{\hbar} \right) \int_{x_1}^{x_2} \left( 2m \left( V(x) - E \right) \right)^{\frac{1}{2}} dx \right]$$

So now you can amaze your friends by calculating the probability that a particle will tunnel through an arbitrary potential. It's the stuff science fiction is made of — well, on the microscopic scale, anyway.

# Particles Unbound: Solving the Schrödinger Equation for Free Particles

What about particles outside any square well — that is, free particles? There are plenty of particles that act freely in the universe, and quantum physics has something to say about them.

Here's the Schrödinger equation:

$$\frac{-\hbar^2}{2m} \frac{d^2}{dx^2} \psi(x) + V(x)\psi(x) = E\psi(x)$$

What if the particle were a free particle, with V(x) = 0? In that case, you'd have the following equation:

$$\frac{-\hbar^2}{2m} \frac{d^2}{dx^2} \psi(x) - E\psi(x) = 0$$

And you can rewrite this as

$$\frac{d^2}{dx^2}\psi(x) + k^2\psi(x) = 0$$

where the wave number, $k$, is $k^2 = \frac{2mE}{\hbar^2}$.

You can write the general solution to this Schrödinger equation as

$$\psi(x) = Ae^{ikx} + Be^{-ikx}$$

If you add time-dependence to the equation, you get this time-dependent wave function:

$$\psi(x,\, t) = A \exp\left(ikx - \frac{i\hbar k^2 t}{2m}\right) + B \exp\left(-\left(ikx - \frac{i\hbar k^2 t}{2m}\right)\right)$$

That's a solution to the Schrödinger equation, but it turns out to be unphysical. To see this, note that for either term in the equation, you can't normalize the probability density, $|\psi(x)|^2$ (see the earlier section titled "Normalizing the wave function" for more on normalizing):

$$|\psi(x)|^2 = |A|^2 \text{ or } |B|^2$$

What's going on here? The probability density for the position of the particle is uniform throughout all $x$! In other words, you can't pin down the particle at all.

This is a result of the form of the time-dependent wave function, which uses an exact value for the wave number, $k$ — and $p = \hbar k$ and $E = \hbar k^2/2m$. So what that equation says is that you know E and $p$ exactly. And if you know $p$ and E exactly, that causes a large uncertainty in $x$ and $t$ — in fact, $x$ and $t$ are completely uncertain. That doesn't correspond to physical reality.

For that matter, the wave function $\psi(x)$, as it stands, isn't something you can normalize. Trying to normalize the first term, for example, gives you this integral:

$$\int_{-\infty}^{+\infty} \psi(x)\psi^*(x)dx$$

And for the first term of $\psi(x,\, t)$, this is

$$\int_{-\infty}^{+\infty} \psi(x)\psi^*(x)dx = |A|^2 \int_{-\infty}^{+\infty} dx \to \infty$$

And the same is true of the second term in $\psi(x, t)$.

So what do you do here to get a physical particle? The next section explains.

## Getting a physical particle with a wave packet

If you have a number of solutions to the Schrödinger equation, any linear combination of those solutions is also a solution. So that's the key to getting a physical particle: You add various wave functions together so that you get a *wave packet*, which is a collection of wave functions of the form $e^{i(kx - Et/\hbar)}$ such that the wave functions interfere constructively at one location and interfere destructively (go to zero) at all other locations:

$$\psi(x, t) = \sum_{n-1}^{\infty} \phi_n e^{i\left(kx - Et/\hbar\right)}$$

This is usually written as a continuous integral:

$$\psi(x, t) = \frac{1}{(2\pi)^{1/2}} \int_{-\infty}^{+\infty} \phi(k, t) e^{i\left(kx - Et/\hbar\right)} dk$$

What is $\phi(k, t)$? It's the amplitude of each component wave function, and you can find $\phi(k, t)$ from the Fourier transform of the equation:

$$\phi(k, t) = \frac{1}{(2\pi)^{1/2}} \int_{-\infty}^{+\infty} \psi(x, t) e^{-i\left(kx - Et/\hbar\right)} dx$$

Because $k = p/\hbar$, you can also write the wave packet equations like this, in terms of $p$, not $k$:

$$\psi(x, t) = \frac{1}{(2\pi)^{1/2}} \int_{-\infty}^{+\infty} \psi(p, t) e^{i\left(px - Et/\hbar\right)} dp$$

$$\phi(k, t) = \frac{1}{(2\pi)^{1/2}} \int_{-\infty}^{+\infty} \psi(x, t) e^{-i\left(kx - Et/\hbar\right)} dx$$

Well, you may be asking yourself just what's going on here. It looks like $\psi(x, t)$ is defined in terms of $\phi(p, t)$, but $\phi(p, t)$ is defined in terms of $\psi(x, t)$. That looks pretty circular.

The answer is that the two previous equations aren't definitions of $\psi(x, t)$ or $\phi(p, t)$; they're just equations relating the two. You're free to choose your own wave packet shape yourself — for example, you may specify the shape of $\phi(p, t)$, and $\psi\left(x, t\right) = \dfrac{1}{\left(2\pi\right)^{1/2}} \displaystyle\int_{-\infty}^{+\infty} \phi\left(k, t\right) e^{i\left(kx - Et/\hbar\right)} dk$ would let you find $\psi(x, t)$.

# Going through a Gaussian example

Here's an example in which you get concrete, selecting an actual wave packet shape. Choose a so-called Gaussian wave packet, which you can see in Figure 3-12 — localized in one place, zero in the others.

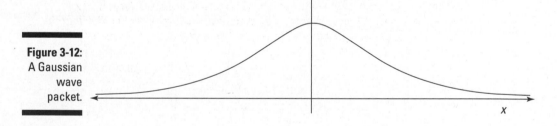

**Figure 3-12:** A Gaussian wave packet.

The amplitude $\phi(k)$ you may choose for this wave packet is

$$\psi\left(k\right) = A \exp\left[-a^2\left(k - k_0\right)\middle/ 4\right]$$

You start by normalizing $\phi(k)$ to determine what A is. Here's how that works:

$$1 = \int_{-\infty}^{+\infty} \left|\phi\left(k\right)\right|^2 dk$$

Substituting in $\phi(k)$ gives you this equation:

$$1 = \left|A\right|^2 \int_{-\infty}^{+\infty} \exp\left[\frac{-a^2}{2}\left(k - k_0\right)^2\right] dk$$

Doing the integral (that means looking it up in math tables) gives you the following:

$$1 = |A|^2 \left[ 2\pi \big/ a^2 \right]^{1/2}$$

Therefore, $A = \left[ a^2 \big/ 2\pi \right]^{1/4}$.

So here's your wave function:

$$\psi(x) = \frac{1}{(2\pi)^{1/2}} \left[ a^2 \big/ 2\pi \right]^{1/4} \int_{-\infty}^{+\infty} \exp\left[ -a^2 (k - k_0) \big/ 4 \right] e^{ikx} dk$$

This little gem of an integral can be evaluated to give you the following:

$$\psi(x) = \left[ 2 \big/ \pi a^2 \right]^{1/4} \exp\left[ -x^2 \big/ a^2 \right] e^{ik_0 x}$$

So that's the wave function for this Gaussian wave packet (**Note:** The $\exp[-x^2/a^2]$ is the Gaussian part that gives the wave packet the distinctive shape that you see in Figure 3-12) — and it's already normalized.

Now you can use this wave packet function to determine the probability that the particle will be in, say, the region $0 \le x \le a/2$. The probability is

$$\int_0^{a/2} |\psi(x)|^2 dx$$

In this case, the integral is

$$\left[ \frac{2}{\pi a^2} \right]^{1/2} \int_0^{a/2} \exp\left( -2x^2 \big/ a^2 \right) dx$$

And this works out to be

$$\left[ 2 \big/ \pi a^2 \right]^{1/2} \int_0^{a/2} \exp\left( -2x^2 \big/ a^2 \right) dx = \frac{1}{3}$$

So the probability that the particle will be in the region $0 \le x \le a/2$ is $1/3$. Cool!

# Chapter 4

# Back and Forth with Harmonic Oscillators

. . . . . . . . . . . . . . . . . . . . . . . . . . . . . . . . . . . . . . . . . . . . . . . . . . . . . . . . .

### In This Chapter

▶ Hamiltonians: Looking at total energy

▶ Solving for energy states with creation and annihilation operators

▶ Understanding the matrix version of harmonic oscillator operators

▶ Writing computer code to solve the Schrödinger equation

. . . . . . . . . . . . . . . . . . . . . . . . . . . . . . . . . . . . . . . . . . . . . . . . . . . . . . . . .

*H*armonic oscillators are physics setups with periodic motion, such as things bouncing on springs or tick-tocking on pendulums. You're probably already familiar with harmonic oscillator problems in the macroscopic arena, but now you're going microscopic. There are many, many physical cases that can be approximated by harmonic oscillators, such as atoms in a crystal structure.

In this chapter, you see both exact solutions to harmonic oscillator problems as well as computational methods for solving them. Knowing how to solve the Schrödinger equation using computers is a useful skill for any quantum physics expert.

## Grappling with the Harmonic Oscillator Hamiltonians

Okay, time to start talking Hamiltonians (and I'm not referring to fans of the U.S. Founding Father Alexander Hamilton). The Hamiltonian will let you find the energy levels of a system.

# *Going classical with harmonic oscillation*

In classical terms, the force on an object in harmonic oscillation is the following (this is Hooke's law):

F = −*kx*

In this equation, *k* is the spring constant, measured in Newtons/meter, and *x* is displacement. The key point here is that the restoring force on whatever is in harmonic motion is proportional to its displacement. In other words, the farther you stretch a spring, the harder it'll pull back.

Because F = *ma,* where *m* is the mass of the particle in harmonic motion and *a* is its instantaneous acceleration, you can substitute for F and write this equation as

*ma* + *kx* = 0

Here's the equation for instantaneous acceleration, where *x* is displacement and *t* is time:

$$a = \frac{d^2 x}{dt^2}$$

So substituting for *a,* you can rewrite the force equation as

$$ma + kx = \frac{md^2 x}{dt^2} + kx = 0$$

Dividing by the mass of the particle gives you the following:

$$\frac{d^2 x}{dt^2} + \frac{kx}{m} = \frac{0}{m}$$

If you take $^k/_m = \omega^2$ (where $\omega$ is the angular frequency), this becomes

$$\frac{d^2 x}{dt^2} + \omega^2 x = 0$$

You can solve this equation for *x,* where A and B are constants:

*x* = A sin$\omega t$ + B cos$\omega t$

Therefore, the solution is an oscillating one because it involves sines and cosines, which represent periodic waveforms.

## Understanding total energy in quantum oscillation

Now look at harmonic oscillators in quantum physics terms. The *Hamiltonian* (H) is the sum of kinetic and potential energies — the total energy of the system:

H = KE + PE

For a harmonic oscillator, here's what these energies are equal to:

✔ The kinetic energy at any one moment is the following, where *p* is the particle's momentum and *m* is its mass:

$$KE = \frac{p^2}{2m}$$

✔ The particle's potential energy is equal to the following, where *k* is the spring constant and *x* is displacement:

$$PE = \frac{1}{2}kx^2 = \frac{1}{2}m\omega^2 x^2$$

(***Note:*** The *k* cancels out because $\omega^2 = {}^k\!/_{m.}$)

Therefore, in quantum physics terms, you can write the Hamiltonian as H = KE + PE, or

$$H = \frac{P^2}{2m} + \frac{1}{2}m\omega^2 X^2$$

where P and X are the momentum and position operators.

You can apply the Hamiltonian operator to various eigenstates (see Chapter 2 for more on eigenstates), $|\psi>$, of the harmonic oscillator to get the total energy, E, of those eigenstates:

$$H|\psi> \frac{P^2}{2m}|\psi> + \frac{1}{2}m\omega^2 X^2|\psi> = E|\psi>$$

The problem now becomes one of finding the eigenstates and eigenvalues. However, this doesn't turn out to be an easy task. Unlike the potentials $V(x)$ covered in Chapter 3, $V(x)$ for a harmonic oscillator is more complex, depending as it does on $x^2$.

So you have to be clever. The way you solve harmonic oscillator problems in quantum physics is with operator algebra — that is, you introduce a new set of operators. And they're coming up now.

# Creation and Annihilation: Introducing the Harmonic Oscillator Operators

Creation and annihilation may sound like big make-or-break the universe kinds of ideas, but they play a starring role in the quantum world when you're working with harmonic oscillators. You use the creation and annihilation operators to solve harmonic oscillator problems because doing so is a clever way of handling the tougher Hamiltonian equation (see the preceding section). Here's what these two new operators do:

- **Creation operator:** The creation operator raises the energy level of an eigenstate by one level, so if the harmonic oscillator is in the fourth energy level, the creation operator raises it to the fifth level.

- **Annihilation operator:** The annihilation operator does the reverse, lowering eigenstates one level.

These operators make it easier to solve for the energy spectrum without a lot of work solving for the actual eigenstates. In other words, you can understand the whole energy spectrum by looking at the energy difference between eigenstates.

## Mind your p's and q's: Getting the energy state equations

Here's how people usually solve for the energy spectrum. First, you introduce two new operators, $p$ and $q$, which are dimensionless; they relate to the P (momentum) operator this way:

- $p = P \Big/ (m\hbar\omega)^{1/2}$

- $q = X \left( m\omega \big/ \hbar \right)^{1/2}$

You use these two new operators, $p$ and $q$, as the basis of the annihilation operator, $a$, and the creation operator, $a^\dagger$:

- ✔ $a = \dfrac{1}{\sqrt{2}}\left(q + ip\right)$

- ✔ $a^\dagger = \dfrac{1}{\sqrt{2}}\left(q - ip\right)$

Now you can write the harmonic oscillator Hamiltonian like this, in terms of $a$ and $a^\dagger$:

$$H = \hbar\omega\left(a^\dagger a + \frac{1}{2}\right)$$

As for creating new operators here, the quantum physicists went crazy, even giving a name to $a^\dagger a$: the $N$ or *number operator.* So here's how you can write the Hamiltonian:

$$H = \hbar\omega\left(N + \frac{1}{2}\right)$$

The N operator returns the *number* of the energy level of the harmonic oscillator. If you denote the eigenstates of N as $|n>$, you get this, where $n$ is the number of the $n$th state:

$N|n> = n|n>$

Because $H = \hbar\omega(N + 1/2)$, and because $H|n> = E_n\,|n>$, then by comparing the previous two equations, you have

$$E_n = \left(n + \frac{1}{2}\right)\hbar\omega \quad n = 0,1,2\ldots$$

Amazingly, that gives you the energy eigenvalues of the $n$th state of a quantum mechanical harmonic oscillator. So here are the energy states:

- ✔ The ground state energy corresponds to $n = 0$:

    $$E_0 = \frac{1}{2}\hbar\omega$$

✔ The first excited state is

$$E_1 = \frac{3}{2}\hbar\omega$$

✔ The second excited state has an energy of

$$E_2 = \frac{5}{2}\hbar\omega$$

And so on. That is, the energy levels are discrete and *nondegenerate* (not shared by any two states). Thus, the energy spectrum is made up of equidistant bands.

# Finding the Eigenstates

When you have the eigenstates (see Chapter 2 to find out all about eigenstates), you can determine the allowable states of a system and the relative probability that the system will be in any of those states.

The commutator of operators A, B is [A, B] = AB – BA, so note that the commutator of $a$ and $a^\dagger$ is the following:

$$\left[a, a^\dagger\right] = \frac{1}{2}\left[q + ip, q - ip\right]$$

This is equal to the following:

$$\left[a, a^\dagger\right] = \frac{1}{2}\left[q + ip, q - ip\right] = -i\left[q, p\right]$$

This equation breaks down to $[a, a^\dagger] = 1$. And putting together this equation with $H = \hbar\omega\left(N + \frac{1}{2}\right)$, you get $\left[a, H\right] = \hbar\omega a$ and $\left[a^\dagger, H\right] = -\hbar\omega a^\dagger$.

### Finding the energy of a|n>

Okay, with the commutator relations, you're ready to go. The first question is if the energy of state $|n>$ is $E_n$, what is the energy of the state $a|n>$? You can write the energy of $a|n>$ this way:

$H(a|n>)$

$= (aH - \hbar\omega a)|n>$

$= (E_n - \hbar\omega)(a|n>)$

So $a|n>$ is also an eigenstate of the harmonic oscillator, with energy $E_n - \hbar\omega$, not $E_n$. That's why $a$ is called the annihilation or lowering operator: It lowers the energy level of a harmonic oscillator eigenstate by one level.

### Finding the energy of $a^\dagger$|n>

So what's the energy level of $a^\dagger|n>$? You can write that can like this:

$H\left(a^\dagger|n>\right)$

$= \left(a^\dagger H + \hbar\omega a^\dagger\right)|n>$

$= \left(E_n + \hbar\omega\right)\left(a^\dagger|n>\right)$

All this means that $a^\dagger|n>$ is an eigenstate of the harmonic oscillator, with energy $E_n + \hbar\omega$, not just $E_n$ — that is, the $a^\dagger$ operator raises the energy level of an eigenstate of the harmonic oscillator by one level.

## Using a and $a^\dagger$ directly

If you've been following along from the preceding section, you know that $H(a|n>) = (E_n - \hbar\omega)(a|n>)$ and $H(a^\dagger|n>) = (E_n + \hbar\omega)(a^\dagger|n>)$. You can derive the following from the these equations:

- $a|n> = C|n-1>$
- $a^\dagger|n> = D|n+1>$

C and D are positive constants, but what do they equal? The states $|n-1>$ and $|n+1>$ have to be *normalized,* which means that $<n-1|n-1> = <n+1|n+1> = 1$. So take a look at the quantity using the C operator:

$$(<n|a^\dagger)(a|n>) = C^2<n-1|n-1>$$

And because $|n-1>$ is normalized, $<n-1|n-1> = 1$:

$$(<n|a^\dagger)(a|n>) = C^2$$
$$<n|a^\dagger a|n> = C^2$$

But you also know that $a^\dagger a = N$, the energy level operator, so you get the following equation:

$$<n|N|n> = C^2$$

$N|n> = n|n>$, where $n$ is the energy level, so

$$n<n|n> = C^2$$

However, $<n|n> = 1$, so

$$n = C^2$$
$$n^{1/2} = C$$

This finally tells you, from $a|n> = C|n-1>$, that

$$a|n> = n^{1/2} |n-1>$$

That's cool — now you know how to use the lowering operator, $a$, on eigenstates of the harmonic oscillator.

What about the raising operator, $a^\dagger$? Following the same course of reasoning you take with the $a$ operator, you can show the following:

$$a^\dagger|n> = (n+1)^{1/2} |n+1>$$

So at this point, you know what the energy eigenvalues are and how the raising and lowering operators affect the harmonic oscillator eigenstates. You've made quite a lot of progress, using the $a$ and $a^\dagger$ operators instead of trying to solve the Schrödinger equation.

# Finding the harmonic oscillator energy eigenstates

The charm of using the operators $a$ and $a^\dagger$ is that given the ground state, $|0>$, those operators let you find all successive energy states. If you want to find an excited state of a harmonic oscillator, you can start with the ground state, $|0>$, and apply the raising operator, $a^\dagger$. For example, you can do this:

✔ $|1> = a^\dagger|0> = |1>$

✔ $|2> = \dfrac{1}{\sqrt{2}}a^\dagger|1> = \dfrac{1}{\sqrt{2!}}\left(a^\dagger\right)^2|0>$

✔ $|3> = \dfrac{1}{\sqrt{3}}a^\dagger|2> = \dfrac{1}{\sqrt{3!}}\left(a^\dagger\right)^3|0>$

✔ $|4> = \dfrac{1}{\sqrt{4}}a^\dagger|3> = \dfrac{1}{\sqrt{4!}}\left(a^\dagger\right)^4|0>$

And so on. In general, you have this relation:

$$|n> = \dfrac{1}{\sqrt{n!}}\left(a^\dagger\right)^n|0>$$

## Working in position space

Okay, $|n> = \dfrac{1}{\sqrt{n!}}\left(a^\dagger\right)^n|0>$ is fine as far as it goes — but just what is $|0>$? Can't you get a spatial eigenstate of this eigenvector? Something like $\psi_0(x)$, not just $|0>$? Yes, you can. In other words, you want to find $<x|0> = \psi_0(x)$. So you need the representations of $a$ and $a^\dagger$ in position space.

The $p$ operator is defined as

$$p = \dfrac{\mathrm{P}}{\left(m\hbar\omega\right)^{1/2}}$$

Because $P = -i\hbar \dfrac{d}{dx}$ , you can write

$$p = \frac{-i\hbar}{(m\hbar\omega)^{\frac{1}{2}}} \frac{d}{dx}$$

And writing $x_0 = [\hbar/(m\hbar\omega)^{1/2}]$, this becomes

$$p = \frac{-i\hbar}{(m\hbar\omega)^{\frac{1}{2}}} \frac{d}{dx} = -ix_0 \frac{d}{dx}$$

Okay, what about the $a$ operator? You know that

$$a = \frac{1}{\sqrt{2}}(q + ip)$$

And that

$$q = X\left(\frac{m\omega}{\hbar}\right)^{\frac{1}{2}} = \frac{X}{x_0}$$

Therefore,

$$a = \frac{1}{\sqrt{2}}\left(\frac{X}{x_0} + x_0 \frac{d}{dx}\right)$$

You can also write this equation as

$$a = \frac{1}{x_0 \sqrt{2}}\left(X + x_0^2 \frac{d}{dx}\right)$$

Okay, so that's $a$ in the position representation. What's $a^\dagger$? That turns out to be this:

$$a^\dagger = \frac{1}{x_0 \sqrt{2}}\left(X - x_0^2 \frac{d}{dx}\right)$$

Now's the time to be clever. You want to solve for |0> in the position space, or <x|0>. Here's the clever part — when you use the lowering operator, *a*, on|0>, you have to get 0 because there's no lower state than the ground state, so *a*|0> = 0. And applying the <x| bra gives you <x|*a*|0> = 0.

That's clever because it's going to give you a homogeneous differential equation (that is, one that equals zero). First, you substitute for *a*:

$$< x|a|0> = 0$$

$$\frac{1}{x_0\sqrt{2}} < x\left|X + x_0^2\frac{d}{dx}\right|0> = 0$$

Then you use $< x|0> = \psi_0(x)$:

$$\frac{1}{x_0\sqrt{2}}\left(\psi_0(x)\right)\left(X + x_0^2\frac{d}{dx}\right) = 0$$

$$\frac{1}{x_0\sqrt{2}}\left(x\psi_0(x) + x_0^2\frac{d\psi_0(x)}{dx}\right) = 0$$

Multiplying both sides by $x_0\sqrt{2}$ gives you the following

$$x\psi_0(x) + x_0^2\frac{d\psi_0(x)}{dx} = 0$$

$$\frac{d\psi_0(x)}{dx} = \frac{-x\psi_0(x)}{x_0^2}$$

The solution to this compact differential equation is

$$\psi_0(x) = A\exp\left(-x^2\Big/2x_0^2\right)$$

That's a gaussian function, so the ground state of a quantum mechanical harmonic oscillator is a gaussian curve, as you see in Figure 4-1.

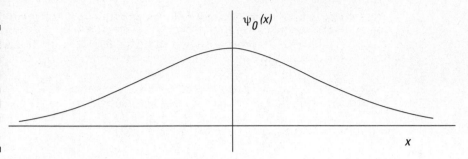

**Figure 4-1:**
The ground
state of a
quantum
mechanical
harmonic
oscillator.

### Finding the wave function of the ground state

As a gaussian curve, the ground state of a quantum oscillator is $\psi_0(x) = A \exp(-x^2/2x_0^2)$. How can you figure out A? Wave functions must be normalized, so the following has to be true:

$$1 = \int_{-\infty}^{\infty} \left| \psi_0(x) \right|^2 dx$$

Substituting for $\psi_0(x)$ gives you this next equation:

$$1 = \int_{-\infty}^{\infty} \left| A \exp\left( -x^2 \big/ 2x_0^{\,2} \right) \right|^2 dx$$

$$1 = A^2 \int_{-\infty}^{\infty} \exp\left( -x^2 \big/ 2x_0^{\,2} \right)^2 dx_0$$

You can evaluate this integral to be

$$1 = A^2 \int_{-\infty}^{\infty} \exp\left( -x^2 \big/ 2x_0^{\,2} \right) dx = A^2 \pi^{\frac{1}{2}} x_0$$

Therefore,

$$1 = A^2 \pi^{\frac{1}{2}} x_0$$

$$A = \frac{1}{\pi^{\frac{1}{4}} x_0^{\frac{1}{2}}}$$

This means that the wave function for the ground state of a quantum mechanical harmonic oscillator is

$$\psi_0(x) = \frac{1}{\pi^{1/4} x_0^{1/2}} \exp\left(-x^2 \Big/ 2x_0^2\right)$$

Cool. Now you've got an exact wave function.

### A little excitement: Finding the first excited state

Okay, the preceding section shows you what $\psi_0(x)$ looks like. What about the first excited state, $\psi_1(x)$? Well, as you know, $\psi_1(x) = <x|1>$ and $|1> = a^\dagger|0>$, so

$$\psi_1(x) = <x|a^\dagger|0>$$

And you know that $a^\dagger$ is the following:

$$a^\dagger = \frac{1}{x_0\sqrt{2}}\left(X - x_0^2 \frac{d}{dx}\right)$$

Therefore, $\psi_1(x) = <x|a^\dagger|0>$ becomes

$$<x|a^\dagger|0> = \frac{1}{x_0\sqrt{2}} <x|\left(X - x_0^2\frac{d}{dx}\right)|0>$$

$$= \frac{1}{x_0\sqrt{2}}\left(X - x_0^2\frac{d}{dx}\right)<x|0>$$

And because $\psi_0(x) = <x|0>$, you get the following equation:

$$\psi_1(x) = \frac{1}{x_0\sqrt{2}} X - x_0^2\frac{d}{dx}\psi_0(x)$$

$$= \frac{1}{x_0\sqrt{2}}\left(x - x_0^2\frac{(-x)}{x_0^2}\right)\psi_0(x)$$

$$= \frac{\sqrt{2}}{x_0} x\psi_0(x)$$

You also know the following:

$$\psi_0(x) = \frac{1}{\pi^{1/4} x_0^{1/2}} \exp\left(-x^2\Big/2x_0^{\,2}\right)$$

Therefore, $\psi_1(x) = \frac{\sqrt{2}}{x_0} x \psi_0(x)$ becomes

$$\psi_1(x) = \frac{\sqrt{2}}{\pi^{1/4} x_0^{3/2}} x \exp\left(-x^2\Big/2x_0^{\,2}\right)$$

What's $\psi_1(x)$ look like? You can see a graph of $\psi_1(x)$ in Figure 4-2, where it has one *node* (transition through the $x$ axis).

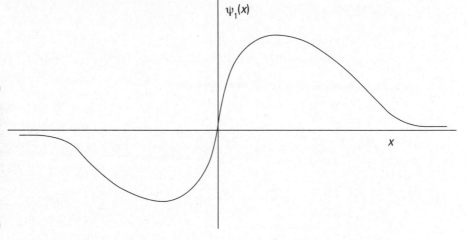

**Figure 4-2:**
The first excited state of a quantum mechanical harmonic oscillator.

### Finding the second excited state

All right, how about finding $\psi_2(x)$ and so on? You can find $\psi_2(x)$ from this equation:

$$\psi_2(x) = \frac{1}{\sqrt{2!}} < x\left|\left(a^\dagger\right)^2\right|0 >$$
$$= \frac{1}{\sqrt{2!}} < x\left|\left(a^\dagger\right)^2\right|0 >$$

Substituting for $a^\dagger$, the equation becomes

$$\psi_2(x) = \frac{1}{\sqrt{2!}} \frac{1}{2!x_0^2} \left( x - x_0^2 \frac{d}{dx} \right)^2 \psi_0(x)$$

## *Using hermite polynomials to find any excited state*

You can generalize the differential equation for $\psi_n(x)$ like this:

$$\psi_n(x) = \frac{1}{\pi^{\frac{1}{4}} (2^n n!)^{\frac{1}{2}} x_0^{n+\frac{1}{2}}} \left( x - x_0^2 \frac{d}{dx} \right)^n \exp\left( -x^2 \Big/ 2x_0^2 \right)$$

To solve this general differential equation, you make use of the fact that

$$\left( x - x_0^2 \frac{d}{dx} \right)^n \exp\left( -x^2 \Big/ 2x_0^2 \right) = \exp\left( -x^2 \Big/ 2x_0^2 \right) H_n\left( x \Big/ x_0 \right)$$

$H_n(x)$ is the $n$th *hermite polynomial*, which is defined this way:

$$H_n(x) = (-1)^n \exp(x^2) \frac{d^n}{dx^n} \exp(-x^2)$$

Holy mackerel! What do the hermite polynomials look like? Here's $H_0(x)$, $H_1(x)$, and so on:

- $H_0(x) = 1$
- $H_1(x) = 2x$
- $H_2(x) = 4x^2 - 2$
- $H_3(x) = 8x^3 - 12x$
- $H_4(x) = 16x^4 - 48x^2 + 12$
- $H_5(x) = 32x^5 - 160x^3 + 120x$

What does this buy you? You can express the wave functions for quantum mechanical harmonic oscillators like this, using the hermite polynomials $H_n(x)$:

$$\psi_n(x) = \frac{1}{\pi^{\frac{1}{4}} (2n!x_0)^{\frac{1}{2}}} H_n\left( x \Big/ x_0 \right) \exp\left( -x^2 \Big/ 2x_0^2 \right)$$

where $x_0 = \left( \hbar \Big/ (m\omega) \right)^{\frac{1}{2}}$.

And that's what the wave function looks like for a quantum mechanical harmonic oscillator. Who knew it would've involved hermite polynomials?

You can see what $\psi_2(x)$ looks like in Figure 4-3; note that there are two nodes here — in general, $\psi_n(x)$ for the harmonic oscillator will have $n$ nodes.

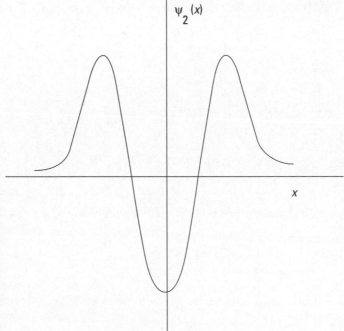

**Figure 4-3:**
The second
excited
state of a
quantum
mechanical
harmonic
oscillator.

## Putting in some numbers

The preceding section gives you $\psi_n(x)$, and you've already solved for $E_n$, so you're on top of harmonic oscillators. Take a look at an example.

Say that you have a proton undergoing harmonic oscillation with $\omega = 4.58 \times 10^{21}$ sec$^{-1}$, as shown in Figure 4-4.

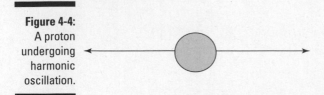

**Figure 4-4:**
A proton
undergoing
harmonic
oscillation.

What are the energies of the various energy levels of the proton? You know that in general,

$$E_n = \left(n + \frac{1}{2}\right)\hbar\omega \quad n = 0,1,2\ldots$$

So here are the energies of the proton, in megaelectron volts (MeV):

✔ $E_0 = \dfrac{\hbar\omega}{2} = 1.50$ MeV

✔ $E_1 = \dfrac{3\hbar\omega}{2} = 4.50$ MeV

✔ $E_2 = \dfrac{5\hbar\omega}{2} = 7.50$ MeV

✔ $E_3 = \dfrac{7\hbar\omega}{2} = 10.50$ MeV

And so on.

Now what about the wave functions? The general form of $\psi_n(x)$ is

$$\psi_n(x) = \frac{1}{\pi^{1/4}\left(2^n n! x_0\right)^{1/2}} H_n\left(\frac{x}{x_0}\right)\exp\left(-\frac{x^2}{2x_0{}^2}\right)$$

where $x_0 = \left(\dfrac{\hbar}{(m\omega)}\right)^{1/2}$. So $x_0 = 3.71 \times 10^{-15}$ m.

Convert all length measurements into femtometers (1 fm = $1 \times 10^{-15}$ m), giving you $x_0 = 3.71$ fm. Here's $\psi_0(x)$, where $x$ is measured in femtometers:

$$\psi_0(x) = \frac{1}{1.92\pi^{1/4}}\exp\left(-\frac{x^2}{27.5}\right)$$

Here are a couple more wave functions:

✔ $\psi_1(x) = \dfrac{1}{2.72\pi^{1/4}}2\left(\dfrac{x}{3.71}\right)\exp\left(-\frac{x^2}{27.5}\right)$

✔ $\psi_2(x) = \dfrac{1}{5.45\pi^{1/4}}\left[4\left(\dfrac{x}{3.71}\right)^2 - 2\right]\exp\left(-\frac{x^2}{27.5}\right)$

# Looking at Harmonic Oscillator Operators as Matrices

Because the harmonic oscillator has regularly spaced energy levels, people often view it in terms of matrices, which can make things simpler. For example, the following may be the ground state eigenvector (note that it's an infinite vector):

$$|0> = \begin{vmatrix} 1 \\ 0 \\ 0 \\ 0 \\ 0 \\ 0 \\ . \\ . \\ . \end{vmatrix}$$

And this may be the first excited state:

$$|1> = \begin{vmatrix} 0 \\ 1 \\ 0 \\ 0 \\ 0 \\ 0 \\ . \\ . \\ . \end{vmatrix}$$

And so on. The N operator, which just returns the energy level, would then look like this:

$$
N =
\begin{vmatrix}
0 & 0 & 0 & 0 & \ldots \\
0 & 1 & 0 & 0 & \ldots \\
0 & 0 & 2 & 0 & \ldots \\
0 & 0 & 0 & 3 & \ldots \\
0 & 0 & 0 & 0 & \ldots \\
0 & 0 & 0 & 0 & \ldots \\
\cdot & \cdot & \cdot & \cdot & \ldots \\
\cdot & \cdot & \cdot & \cdot & \ldots \\
\cdot & \cdot & \cdot & \cdot & \ldots
\end{vmatrix}
$$

So N|2> gives you

$$
N\left|2\right> =
\begin{vmatrix}
0 & 0 & 0 & 0 & \ldots \\
0 & 1 & 0 & 0 & \ldots \\
0 & 0 & 2 & 0 & \ldots \\
0 & 0 & 0 & 3 & \ldots \\
0 & 0 & 0 & 0 & \ldots \\
0 & 0 & 0 & 0 & \ldots \\
\cdot & \cdot & \cdot & \cdot & \ldots \\
\cdot & \cdot & \cdot & \cdot & \ldots \\
\cdot & \cdot & \cdot & \cdot & \ldots
\end{vmatrix}
\begin{vmatrix}
0 \\
0 \\
1 \\
0 \\
0 \\
0 \\
\cdot \\
\cdot \\
\cdot
\end{vmatrix}
$$

This is equal to

$$N|2> =$$

$$
\begin{vmatrix} 0 \\ 0 \\ 2 \\ 0 \\ 0 \\ 0 \\ 0 \\ 0 \\ 0 \end{vmatrix}
=
\begin{vmatrix}
0 & 0 & 0 & 0 & \ldots \\
0 & 1 & 0 & 0 & \ldots \\
0 & 0 & 2 & 0 & \ldots \\
0 & 0 & 0 & 3 & \ldots \\
0 & 0 & 0 & 0 & \ldots \\
0 & 0 & 0 & 0 & \ldots \\
. & . & . & . & \ldots \\
. & . & . & . & \ldots \\
. & . & . & . & \ldots
\end{vmatrix}
\begin{vmatrix} 0 \\ 0 \\ 1 \\ 0 \\ 0 \\ 0 \\ . \\ . \\ . \end{vmatrix}
$$

In other words, $N|2> = 2|2>$.

How about the $a$ (lowering) operator? That looks like this:

$$a =$$

$$
\begin{vmatrix}
0 & \sqrt{1} & 0 & 0 & \ldots \\
0 & 0 & \sqrt{2} & 0 & \ldots \\
0 & 0 & 0 & \sqrt{3} & \ldots \\
0 & 0 & 0 & 0 & \ldots \\
0 & 0 & 0 & 0 & \ldots \\
0 & 0 & 0 & 0 & \ldots \\
. & . & . & . & \ldots \\
. & . & . & . & \ldots \\
. & . & . & . & \ldots
\end{vmatrix}
$$

In this representation, what is $a|1>$? In general, $a|n> = n^{1/2}|n-1>$, so $a|1>$ should equal $|0>$. Take a look:

$$a|1> =$$

$$
\begin{vmatrix}
0 & \sqrt{1} & 0 & 0 & \cdots \\
0 & 0 & \sqrt{2} & 0 & \cdots \\
0 & 0 & 0 & \sqrt{3} & \cdots \\
0 & 0 & 0 & 0 & \cdots \\
0 & 0 & 0 & 0 & \cdots \\
0 & 0 & 0 & 0 & \cdots \\
\cdot & \cdot & \cdot & \cdot & \cdots \\
\cdot & \cdot & \cdot & \cdot & \cdots \\
\cdot & \cdot & \cdot & \cdot & \cdots
\end{vmatrix}
\begin{vmatrix}
0 \\ 1 \\ 0 \\ 0 \\ 0 \\ 0 \\ \cdot \\ \cdot \\ \cdot
\end{vmatrix}
$$

This matrix multiplication equals the following:

$$a|1> =$$

$$
\begin{vmatrix}
\sqrt{1} \\ 0 \\ 0 \\ 0 \\ 0 \\ 0 \\ \cdot \\ \cdot \\ \cdot
\end{vmatrix}
=
\begin{vmatrix}
0 & \sqrt{1} & 0 & 0 & \cdots \\
0 & 0 & \sqrt{2} & 0 & \cdots \\
0 & 0 & 0 & \sqrt{3} & \cdots \\
0 & 0 & 0 & 0 & \cdots \\
0 & 0 & 0 & 0 & \cdots \\
0 & 0 & 0 & 0 & \cdots \\
\cdot & \cdot & \cdot & \cdot & \cdots \\
\cdot & \cdot & \cdot & \cdot & \cdots \\
\cdot & \cdot & \cdot & \cdot & \cdots
\end{vmatrix}
\begin{vmatrix}
0 \\ 1 \\ 0 \\ 0 \\ 0 \\ 0 \\ \cdot \\ \cdot \\ \cdot
\end{vmatrix}
$$

In other words, $a|1> = |0>$, just as expected.

So how about the $a^\dagger$ (raising) operator? Here's how it works in general: $a^\dagger|n> = (n + 1)^{1/2}|n + 1>$. In matrix terms, $a^\dagger$ looks like this:

$$a^\dagger = \begin{vmatrix} 0 & 0 & 0 & 0 & \ldots \\ \sqrt{1} & 0 & 0 & 0 & \ldots \\ 0 & \sqrt{2} & 0 & 0 & \ldots \\ 0 & 0 & \sqrt{3} & 0 & \ldots \\ 0 & 0 & 0 & \sqrt{4} & \ldots \\ 0 & 0 & 0 & 0 & \ldots \\ . & . & . & . & \ldots \\ . & . & . & . & \ldots \\ . & . & . & . & \ldots \end{vmatrix}$$

For example, you expect that $a^\dagger|1> = \sqrt{2}\,|1>$. Does it? The matrix multiplication is

$$a^\dagger|1> = \begin{vmatrix} 0 & 0 & 0 & 0 & \ldots \\ \sqrt{1} & 0 & 0 & 0 & \ldots \\ 0 & \sqrt{2} & 0 & 0 & \ldots \\ 0 & 0 & \sqrt{3} & 0 & \ldots \\ 0 & 0 & 0 & \sqrt{4} & \ldots \\ 0 & 0 & 0 & 0 & \ldots \\ . & . & . & . & \ldots \\ . & . & . & . & \ldots \\ . & . & . & . & \ldots \end{vmatrix} \begin{vmatrix} 0 \\ 1 \\ 0 \\ 0 \\ 0 \\ 0 \\ . \\ . \\ . \end{vmatrix}$$

This equals the following:

$$a^\dagger |1> =$$

$$
\begin{vmatrix} 0 \\ 0 \\ \sqrt{2} \\ 0 \\ 0 \\ 0 \\ \cdot \\ \cdot \\ \cdot \end{vmatrix} =
\begin{vmatrix} 0 & 0 & 0 & 0 & \dots \\ \sqrt{1} & 0 & 0 & 0 & \dots \\ 0 & \sqrt{2} & 0 & 0 & \dots \\ 0 & 0 & \sqrt{3} & 0 & \dots \\ 0 & 0 & 0 & \sqrt{4} & \dots \\ 0 & 0 & 0 & 0 & \dots \\ \cdot & \cdot & \cdot & \cdot & \dots \\ \cdot & \cdot & \cdot & \cdot & \dots \\ \cdot & \cdot & \cdot & \cdot & \dots \end{vmatrix}
\begin{vmatrix} 0 \\ 1 \\ 0 \\ 0 \\ 0 \\ 0 \\ \cdot \\ \cdot \\ \cdot \end{vmatrix}
$$

So $a^\dagger |1> = \sqrt{2}\ |2>$, as it should.

How about taking a look at the Hamiltonian, which returns the energy of an eigenstate, $H|n> = E_n|n>$? In matrix form, the Hamiltonian looks like this:

$$H =$$

$$
\frac{\hbar\omega}{2}
\begin{vmatrix} 1 & 0 & 0 & 0 & \dots \\ 0 & 3 & 0 & 0 & \dots \\ 0 & 0 & 5 & 0 & \dots \\ 0 & 0 & 0 & 7 & \dots \\ 0 & 0 & 0 & 0 & \dots \\ 0 & 0 & 0 & 0 & \dots \\ \cdot & \cdot & \cdot & \cdot & \dots \\ \cdot & \cdot & \cdot & \cdot & \dots \\ \cdot & \cdot & \cdot & \cdot & \dots \end{vmatrix}
$$

So if you prefer the matrix way of looking at things, that's how it works for the harmonic oscillator.

# A Jolt of Java: Using Code to Solve the Schrödinger Equation Numerically

Here's the one-dimensional Schrödinger equation:

$$\frac{-\hbar^2}{2m}\frac{d^2\psi(x)}{dx^2}+V(x)\psi(x)=E\psi(x)$$

And for harmonic oscillators, you can write the equation like this, where $k^2=2m\left(\frac{E-V(x)}{\hbar^2}\right)$:

$$\frac{d^2\psi(x)}{dx^2}+k^2\psi(x)=0$$

In general, as the potential $V(x)$ gets more and more complex, using a computer to solve the Schrödinger equation begins to look more and more attractive. In this section, I show you how to do just that for the harmonic oscillator Schrödinger equation.

## Making your approximations

In computer terms, you can approximate $\psi(x)$ as a collection of points, $\psi_1$, $\psi_2$, $\psi_3$, $\psi_4$, $\psi_5$, and so on, as you see in Figure 4-5.

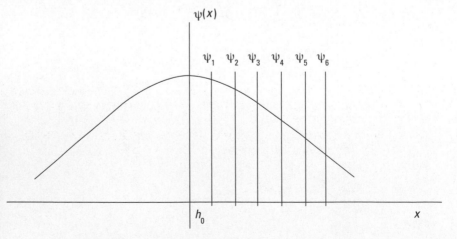

**Figure 4-5:**
Dividing $\psi(x)$ along the $x$ axis.

Each point along $\psi(x)$ — $\psi_1$, $\psi_2$, $\psi_3$, $\psi_4$, $\psi_5$, and so on — is separated from its neighbor by a distance, $h_0$, along the $x$ axis. And because $d\psi/dx$ is the slope of $\psi(x)$, you can make the approximation that

$$\frac{d\psi}{dx} \approx \frac{\psi_{n+1} - \psi_n}{h_0}$$

In other words, the slope, $d\psi/dx$, is approximately equal to $\Delta y/\Delta x$, which is equal to $\psi_{n+1} - \psi_n$ (= $\Delta y$) divided by $h_0$ (= $\Delta x$).

You can rearrange the equation to this:

$$\psi_{n+1} \approx h_0 \frac{d\psi}{dx} + \psi_n$$

That's a crude approximation for $\psi_{n+1}$, given $\psi_n$. So, for example, if you know $\psi_4$, you can find the approximate value of $\psi_5$, if you know $d\psi/dx$ in the region of $\psi_4$.

You can, of course, find better approximations for $\psi_{n+1}$. In particular, physicists often use the *Numerov algorithm* when solving the Schrödinger equation, and that algorithm gives you $\psi_{n+1}$ in terms of $\psi_n$ and $\psi_{n-1}$. Here's what the Numerov algorithm says:

$$\psi_{n+1} = \frac{2\left(1 - \left(5h_0{}^2 k_n(x)^2 \middle/ 12\right)\right)\psi_n - \left(1 + \left(h_0{}^2 k_{n-1}(x)^2\right) \middle/ 12\right)\psi_{n-1}}{1 + h_0{}^2 k_{n+1}(x)^2 \middle/ 12}$$

In this equation, $k_n{}^2(x) = \dfrac{2m\left(E_n - m\omega^2 x^2 \middle/ 2\right)}{\hbar^2}$

and the boundary conditions are $\psi(-\infty) = \psi(\infty) = 0$. Wow. Imagine having to calculate this by hand. Why not leave it up to the computer?

For a proton undergoing harmonic oscillation with $\omega = 4.58 \times 10^{21} \text{ sec}^{-1}$, the exact ground state energy is

$$E_0 = \frac{\hbar\omega}{2} = 1.50 \text{ MeV}$$

You solve this problem computationally earlier in this chapter. The following sections have you try to get this same result using the Numerov algorithm and a computer.

## Building the actual code

To calculate the ground state energy of the harmonic oscillator using the Numerov algorithm, this section uses the Java programming language, which you can get for free from java.sun.com.

Here's how you use the program: You choose a trial value of the energy for the ground state, $E_0$, and then calculate $\psi(x)$ at $\infty$, which should be zero — and if it's not, you can adjust your guess for $E_0$ and try again. You keep going until $\psi(\infty) = 0$ (or if not actually 0, a very small number in computer terms) — and when it does, you know you've guessed the correct energy.

### Approximating $\psi(\infty)$

How do you calculate $\psi(\infty)$? After all, infinity is a pretty big number, and the computer is going to have trouble with that. In practical terms, you have to use a number that approximates infinity. In this case, you can use the classical turning points of the proton — the points where all the proton's energy is potential energy and it has stopped moving in preparation for reversing its direction.

At the turning points, $x_t$, $E_0 = \dfrac{m\omega^2 x^2}{2}$ (that is, all the energy is in potential energy), so

$$x_t = \left( 2E_0 \middle/ \left( m\omega^2 \right) \right)^{\!1/2}$$

And this is on the order of $\pm 5$ femtometers (fm), so you assume that $\psi(x)$ should surely be zero at, say, $\pm 15$ fm. Here's the interval over which you calculate $\psi(x)$:

- $x_{min} = -15$ fm
- $x_{max} = 15$ fm

Divide this 30 fm interval into 200 segments, making the width of each segment, $h_0$, equal to $(x_{max} - x_{min})/200 = h_0 = 0.15$ fm.

Okay, you're making progress. You'll start by assuming that $\psi(x_{min}) = 0$, guess a value of $E_0$, and then calculate $\psi(x_{max}) = \psi_{200}$ (because there are 200 segments, at $x = x_{max}$, $\psi_n = \psi_{200}$), which should equal zero when you get $E_0$.

Here's what the results tell you:

- **Correct:** If abs($\psi_{200}$) is zero, or in practical terms, less than, say, your maximum allowed value of $\psi_{max} = 1 \times 10^{-8}$, then you're done — the $E_0$ you guessed was correct.

- **Too high:** If abs($\psi_{200}$) is larger than your maximum allowed $\psi$, $\psi_{max}$ (= $1 \times 10^{-8}$), and $\psi_{200}$ is *positive*, the energy you chose for $E_0$ was too high. You have to subtract a small amount of energy, $\Delta E$ — say $1 \times 10^{-7}$ MeV — from your guess for the energy; then calculate abs($\psi_{200}$) again and see whether it's still higher than your maximum allowed $\psi$, $\psi_{max}$. If so, you have to repeat the process again.

- **Too low:** If abs($\psi_{200}$) is larger than your maximum allowed $\psi$, $\psi_{max}$ (= $1 \times 10^{-8}$), and $\psi_{200}$ is *negative*, the energy you chose for $E_0$ was too low. You have to add a small amount of energy, $\Delta E$, to your guess for the energy; then calculate abs($\psi_{200}$) again and see whether it's still higher than your maximum allowed $\psi$, $\psi_{max}$. If so, you have to repeat the process.

So how do you calculate $\psi_{200}$? Given two starting values, $\psi_0$ and $\psi_1$, use the Numerov algorithm:

$$\psi_{n+1} = \frac{2\left(1 - \left(5h_0^2 k_n(x)^2 \middle/ 12\right)\right)\psi_n - \left(1 + \left(h_0^2 k_{n-1}(x)^2\right) \middle/ 12\right)\psi_{n-1}}{1 + h_0^2 k_{n+1}(x)^2 \middle/ 12}$$

Keep calculating successive points along $\psi(x)$: $\psi_2$, $\psi_3$, $\psi_4$, and so on. The last point is $\psi_{200}$.

Okay, you're on our way. You're going to start the code with the assumption that $\psi_0 = 0$ and $\psi_1$ is a very small number (you can choose any small number you like). Because you know that the exact ground level energy is actually 1.50 MeV, start the code with the guess that $E_0 = 1.4900000$ MeV and let the computer calculate the actual value using increments of $\Delta E = 1 \times 10^{-7}$ MeV.

Note also this equation depends on $k_n(x)^2$, $k_{n-1}(x)^2$, and $k_{n+1}(x)^2$. Here's how you can find these values, where $E_{current}$ is the current guess for the ground state energy (substitute $n$, $n-1$, and $n+1$ for $j$):

$$k_j^2(x_j) = 2m\left(E_{current} - m\omega^2 x_j^2 \middle/ 2\right) \hbar^2$$

And you know that K = $4.58 \times 10^{21}$ sec$^{-1}$, so

> ✔ $\dfrac{m^2 \omega^2}{\hbar^2} = 5.63 \times 10^{-3}$ fm$^{-4}$
>
> ✔ $\dfrac{2m}{\hbar^2} = 0.05$ MeV$^{-1}$fm$^{-2}$

Therefore, $k_j^2(x_j) = 0.05 E_{current} - 5.63 \times 10^{-3} x_j^2$, where $x_j$ for a particular segment $j$ is $x_j = jh_0 + x_{min}$.

## Writing the code

Okay, now I'm going to put together all the info from the preceding section into some Java code. You start with a Java class, *se* (for Schrödinger Equation), in a file you name *se.java:*

```
public class se
    .
    .
    .
}
```

Then you set up the variables and constants you'll need, including an array for the values you calculate for $\psi$ (because to find $\psi_{n+1}$, you'll have had to store the already-calculated values of $\psi_n$ and $\psi_{n-1}$):

```
public class se
{

  double psi[];
  double ECurrent;
  double Emin = 1.490;
  double xMin = -15.;
  double xMax = 15.;
  double hZero;
  double EDelta = 0.0000001;
  double maxPsi = 0.00000001;
  int numberDivisions = 200;

     .
     .
     .

}
```

The *se* class's constructor gets run first, so you initialize values there, including $\psi_0$ (= $\psi(x_{min})$ = 0) and $\psi_1$ (any small number you want) to get the calculation going. In the main method, called *after the constructor,* you create an object of the *se* class and call it *calculate method* to get things started:

```
public class se
{

  double psi[];
  double ECurrent;
  double Emin = 1.490;
  double xMin = -15.;
  double xMax = 15.;
  double hZero;
  double EDelta = 0.0000001;
  double maxPsi = 0.00000001;
  int numberDivisions = 200;

  public se()
  {
    ECurrent = Emin;
    psi = new double[numberDivisions + 1];
    psi[0] = 0;
    psi[1] = -0.000000001;
    psi[numberDivisions] = 1.0;
    hZero = (xMax - xMin) / numberDivisions;
  }

  public static void main(String [] argv)
  {
    se de = new se();
    de.calculate();
  }

        .
        .
        .

}
```

The real work takes place in the *calculate method,* where you use the current guess for the energy and calculate $\psi_{200}$:

✔ If abs($\psi_{200}$) is less than your maximum allowed value of $\psi$, $\psi_{max}$, you've found the answer — your current guess for the energy is right on.

✔ If abs($\psi_{200}$) is greater than $\psi_{max}$ and $\psi_{200}$ is positive, you have to subtract $\Delta E$ from your current guess for the energy and try again.

✔ If abs($\psi_{200}$) is greater than $\psi_{max}$ and $\psi_{200}$ is negative, you have to add $\Delta E$ to your current guess for the energy and then try again.

Here's what all this looks like in code:

```
public void calculate()
{
  while(Math.abs(psi[numberDivisions])> maxPsi){
    for (int i = 1; i <numberDivisions; i++){
      psi[i + 1] = calculateNextPsi(i);
    }
    if (psi[numberDivisions]> 0.0) {
      ECurrent = ECurrent ñ EDelta;
    }
    else {
      ECurrent = ECurrent + EDelta;
    }
    System.out.println(ìPsi200: ì + psi[numberDivisions]
          + ì E: ì + round(ECurrent));
  }
  System.out.println(ì\nThe ground state energy is ì +
          round(ECurrent) + ì MeV.î);
}
```

Note that the next value of $\psi$ (that is, $\psi_{n+1}$) is calculated with a method named *calculateNextPsi*. Here's where you use the Numerov algorithm — given $\psi_n$, $\psi_{n-1}$, you can calculate $\psi_{n+1}$:

```
public double calculateNextPsi(int n)
{
  double KSqNMinusOne = calculateKSquared(n - 1);
  double KSqN = calculateKSquared(n);
  double KSqNPlusOne = calculateKSquared(n + 1);
  double nextPsi = 2.0 *(1.0 - (5.0 * hZero * hZero *
        KSqN / 12.0)) * psi[n];
  nextPsi = nextPsi - (1.0 + (hZero * hZero *
        KSqNMinusOne / 12.0)) * psi[n - 1];
  nextPsi = nextPsi /(1.0 + (hZero * hZero * KSqNPlusOne
        / 12.0));
  return nextPsi;
}
```

Finally, note that to calculate $\psi_{n+1}$, you need $k_n$, $k_{n-1}$, and $k_{n+1}$, which you find with a method named *calculateKSquared,* which uses the numeric values you've already figured out for this problem:

```
public double calculateKSquared(int n)
  {
    double x = (hZero * n) + xMin;
    return (((0.05) * ECurrent) - ((x * x) * 5.63e-3));
  }
```

Whew. Here's the whole program, *se.java:*

```
public class se
{

  double psi[];
  double ECurrent;
  double Emin = 1.490;
  double xMin = -15.;
  double xMax = 15.;
  double hZero;
  double EDelta = 0.0000001;
  double maxPsi = 0.00000001;
  int numberDivisions = 200;

  public se()
  {
    ECurrent = Emin;
    psi = new double[numberDivisions + 1];
    psi[0] = 0;
    psi[1] = -0.000000001;
    psi[numberDivisions] = 1.0;
    hZero = (xMax - xMin) / numberDivisions;
  }

  public static void main(String [] argv)
  {
    se de = new se();
    de.calculate();
  }

  public void calculate()
  {
    while(Math.abs(psi[numberDivisions])> maxPsi){
      for (int i = 1; i <numberDivisions; i++){
        psi[i + 1] = calculateNextPsi(i);
      }
      if (psi[numberDivisions]> 0.0) {
        ECurrent = ECurrent - EDelta;
      }
      else {
        ECurrent = ECurrent + EDelta;
      }
      System.out.println(ìPsi200: ì + psi[numberDivisions]
          + ì E: ì + round(ECurrent));
    }
    System.out.println(ì\nThe ground state energy is ì +
          round(ECurrent) + ì MeV.î);
```

```
    }

    public double calculateKSquared(int n)
    {
      double x = (hZero * n) + xMin;
      return (((0.05) * ECurrent) - ((x * x) * 5.63e-3));
    }

    public double calculateNextPsi(int n)
    {
      double KSqNMinusOne = calculateKSquared(n - 1);
      double KSqN = calculateKSquared(n);
      double KSqNPlusOne = calculateKSquared(n + 1);
      double nextPsi = 2.0 *(1.0 - (5.0 * hZero * hZero *
            KSqN / 12.0)) * psi[n];
      nextPsi = nextPsi - (1.0 + (hZero * hZero *
            KSqNMinusOne / 12.0)) * psi[n - 1];
      nextPsi = nextPsi /(1.0 + (hZero * hZero * KSqNPlusOne
            / 12.0));
      return nextPsi;
    }

    public double round(double val)
    {
      double divider = 100000;
      val = val * divider;
      double temp = Math.round(val);
      return (double)temp / divider;
    }
}
```

Okay, now you can compile the code with javac, the Java compiler (if javac isn't in your computer's path, be sure to add the correct path to your command-line command, such as `C:>C:\java\bin\javac se.java`).

```
C:>javac se.java
```

This creates *se.class* from *se.java,* and you can run *se.class* with Java itself (adding the correct path if needed):

```
C:>java se
```

## Running the code

When you run the java code for the harmonic oscillator Schrödinger equation, it displays the successive values of $\psi_{200}$ as it adjusts the current guess for the energy as it narrows in on the right answer — which it displays at the end of the run. Here's what you see:

```
C:>java se
PSI200: -1.0503644097337778E-4 E: 1.49
PSI200: -1.050354423295303E-4 E: 1.49
PSI200: -1.0503444368533108E-4 E: 1.49
PSI200: -1.0503344504260495E-4 E: 1.49
              .
              .
              .
PSI200: -6.12820872814324E-8 E: 1.50066
PSI200: -6.031127521356655E-8 E: 1.50066
PSI200: -5.934046348307554E-8 E: 1.50066
PSI200: -5.836965180600015E-8 E: 1.50066
PSI200: -5.739883979461778E-8 E: 1.50066
PSI200: -5.6428029151212084E-8 E: 1.50066
PSI200: -5.5457218252899224E-8 E: 1.50066
PSI200: -5.4486408066519986E-8 E: 1.50066
PSI200: -5.351559702201636E-8 E: 1.50066
PSI200: -5.254478723976338E-8 E: 1.50066
PSI200: -5.157397714326237E-8 E: 1.50066
PSI200: -5.060316801012202E-8 E: 1.50066
PSI200: -4.963235841725704E-8 E: 1.50066
PSI200: -4.866154915227413E-8 E: 1.50066
PSI200: -4.769074041927121E-8 E: 1.50066
PSI200: -4.6719932089691944E-8 E: 1.50066
PSI200: -4.574912368974434E-8 E: 1.50066
PSI200: -4.4778315322587505E-8 E: 1.50066
PSI200: -4.380750790476514E-8 E: 1.50066
PSI200: -4.28367005783992E-8 E: 1.50066
PSI200: -4.186589345217578E-8 E: 1.50066
PSI200: -4.0895085873184064E-8 E: 1.50066
PSI200: -3.992427935226201E-8 E: 1.50066
PSI200: -3.8953472673066213E-8 E: 1.50066
PSI200: -3.79826665057731E-8 E: 1.50066
PSI200: -3.701186038502826E-8 E: 1.50066
PSI200: -3.604105453620266E-8 E: 1.50066
PSI200: -3.507024949509914E-8 E: 1.50066
PSI200: -3.4099444217875174E-8 E: 1.50066
PSI200: -3.312863911389194E-8 E: 1.50066
PSI200: -3.2157834719961815E-8 E: 1.50066
PSI200: -3.1187030089902856E-8 E: 1.50066
PSI200: -3.021622619594536E-8 E: 1.50066
```

```
PSI200: -2.9245421985136167E-8 E: 1.50066
PSI200: -2.8274618172375295E-8 E: 1.50066
PSI200: -2.7303815344369703E-8 E: 1.50066
PSI200: -2.633301196069577E-8 E: 1.50066
PSI200: -2.5362208888510866E-8 E: 1.50066
PSI200: -2.439140632085814E-8 E: 1.50066
PSI200: -2.342060424823075E-8 E: 1.50066
PSI200: -2.244980221960756E-8 E: 1.50066
PSI200: -2.147900005347249E-8 E: 1.50067
PSI200: -2.0508198285622532E-8 E: 1.50067
PSI200: -1.9537397616823192E-8 E: 1.50067
PSI200: -1.8566596602866105E-8 E: 1.50067
PSI200: -1.7595795286272332E-8 E: 1.50067
PSI200: -1.6624994703779555E-8 E: 1.50067
PSI200: -1.565419461892862E-8 E: 1.50067
PSI200: -1.4683394780836424E-8 E: 1.50067
PSI200: -1.3712594592034165E-8 E: 1.50067
PSI200: -1.2741795159638587E-8 E: 1.50067
PSI200: -1.177099622966848E-8 E: 1.50067
PSI200: -1.0800197142733883E-8 E: 1.50067
PSI200: -9.82939798529632E-9 E: 1.50067

The ground state energy is 1.50067 MeV.
```

And there you have it — the program approximates the ground state energy as 1.50067 MeV, pretty darn close to the value you calculated theoretically before: 1.50 MeV.

Very cool.

# Part III
# Turning to Angular Momentum and Spin

The 5th Wave                    By Rich Tennant

©RICHTENNANT

"Great-differential equations brought us
Newton's Law of Universal Gravitation,
Maxwell's field equations, and now Stuart's
Rate of Hair Loss."

# In this part . . .

Things that spin and rotate — that's the topic of this part. Quantum physics has all kinds of things to say about how angular momentum and spin are quantized, and you see it all in this part.

# Chapter 5

# Working with Angular Momentum on the Quantum Level

*I*n classical mechanics, you may measure angular momentum by attaching a golf ball to a string and whirling it over your head. In quantum mechanics, think in terms of a single molecule made up of two bound atoms rotating around each other. That's the level at which quantum mechanical effects become noticeable. And at that level, it turns out that angular momentum is quantized. And since that has tangible results in many cases, such as the spectrum of excited atoms, it's an important topic.

Besides having kinetic and potential energy, particles can also have *rotational energy*. Here's what the Hamiltonian (total energy — see Chapter 4) looks like:

$$H = \frac{L^2}{2I}$$

Here, L is the angular momentum operator and I is the rotation moment of inertia. What are the eigenstates of angular momentum? If L is the angular momentum operator, and *l* is an eigenvalue of L, then you could write the following:

$$H|l> = \frac{L^2}{2I}|l> \quad \text{Incomplete!}$$

But that turns out to be incomplete because angular momentum is a vector in three-dimensional space — and it can be pointing any direction. Angular momentum is typically given by a magnitude and a component in one direction, which is usually the Z direction. So in addition to the magnitude $l$, you also specify the component of L in the Z direction, $L_z$ (the choice of Z is arbitrary — you can just as easily use the X or Y direction).

If the quantum number of the Z component of the angular momentum is designated by $m$, then the complete eigenstate is given by $|l, m>$, so the equation becomes the following:

$$H|l,m> = \frac{L^2}{2I}|l,m>$$

That's the kind of discussion about eigenstates that I cover in this chapter, and I begin with a discussion of angular momentum.

# Ringing the Operators: Round and Round with Angular Momentum

Take a look at Figure 5-1, which depicts a disk rotating in 3D space. Because you're working in 3D, you have to go with vectors to represent both magnitude and direction.

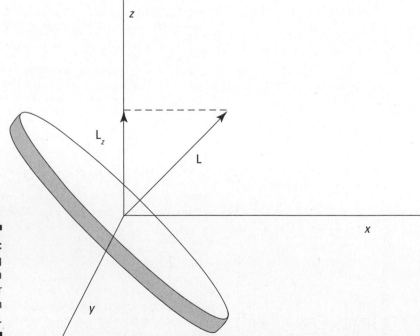

**Figure 5-1:**
A rotating disk with angular momentum vector L.

As you can see, the disk's angular momentum vector, **L,** points perpendicular to the plane of rotation. Here, you can apply the right-hand rule: If you wrap your right hand in the direction something is rotating, your thumb points in the direction of the **L** vector.

Having the **L** vector point out of the plane of rotation has some advantages. For example, if something is rotating at a constant angular speed, the **L** vector will be constant in magnitude and direction — which makes more sense than having the **L** vector rotating in the plane of the disk's rotation and constantly changing direction.

Because **L** is a 3D vector, it can point in any direction, which means that it has $x$, $y$, and $z$ components, $L_x$, $L_y$, and $L_z$ (which aren't vectors, just magnitudes). You can see $L_z$ in Figure 5-1.

**L** is the vector product of **R** (position) and **P** (**L** = **R** × **P**). You can also write $L_x$, $L_y$, and $L_z$ at any given moment in terms of operators like this, where $P_x$, $P_y$, and $P_z$ are the *momentum operators* (which return the momentum in the $x$, $y$, and $z$ directions) and X, Y, and Z are the *position operators* (which return the position in the $x$, $y$, and $z$ directions):

✔ $L_x = YP_z - ZP_y$

✔ $L_y = ZP_x - XP_z$

✔ $L_z = XP_y - YP_x$

You can write the momentum operators $P_x$, $P_y$, and $P_z$ as

$$P_x = -i\hbar \frac{\partial}{\partial x}$$

$$P_y = -i\hbar \frac{\partial}{\partial y}$$

$$P_z = -i\hbar \frac{\partial}{\partial z}$$

Therefore, substituting these operators in the $L_x$, $L_y$, and $L_z$ equations, you can write the equations as

✔ $L_x = -i\hbar \left( Y\frac{\partial}{\partial z} - Z\frac{\partial}{\partial y} \right)$

✔ $L_y = -i\hbar \left( Z\frac{\partial}{\partial x} - X\frac{\partial}{\partial z} \right)$

✔ $L_z = -i\hbar \left( X\frac{\partial}{\partial y} - Y\frac{\partial}{\partial x} \right)$

# *Finding Commutators of L$_x$, L$_y$, and L$_z$*

First examine L$_x$, L$_y$, and L$_z$ by taking a look at how they commute; if they commute (for example, if [L$_x$, L$_y$] = 0), then you can measure any two of them (L$_x$ and L$_y$, for example) exactly. If not, then they're subject to the *uncertainty relation,* and you can't measure them simultaneously exactly.

Okay, so what's the commutator of L$_x$ and L$_y$? Using L$_x$ = YP$_z$ – ZP$_y$ and L$_y$ = ZP$_x$ – XP$_z$, you can write the following:

$$[L_x, L_y] = [YP_z - ZP_y, ZP_x - XP_z]$$

You can write this equation as

$$[L_x, L_y] = [YP_z, ZP_x] - [YP_z, XP_z] - [ZP_y, ZP_x] + [ZP_y, XP_z]$$
$$= Y[P_z, ZP_x]P_x + X[Z, P_{yz}]P_y$$
$$= i\hbar(XP_y - YP_x)$$

But XP$_y$ – YP$_x$ = L$_z$, so [L$_x$, L$_y$] = $i\hbar$L$_z$. So L$_x$ and L$_y$ don't commute, which means that you can't measure them both simultaneously with complete precision. You can also show that [L$_y$, L$_z$] = $i\hbar$L$_x$ and [L$_z$, L$_x$] = $i\hbar$L$_y$.

Because none of the components of angular momentum commute with each other, you can't measure any two simultaneously with complete precision. Rats.

That also means that the L$_x$, L$_y$, and L$_z$ operators can't share the same eigenstates. So what can you do? How can you find an operator that shares eigenstates with the various components of L so that you can write the eigenstates as |l, m>?

The usual trick here is that the square of the angular momentum, L$^2$, is a scalar, not a vector, so it'll commute with the L$_x$, L$_y$, and L$_z$ operators, no problem:

    ✔ [L$^2$, L$_x$] = 0

    ✔ [L$^2$, L$_y$] = 0

    ✔ [L$^2$, L$_z$] = 0

Okay, cool, you're making progress. Because L$_x$, L$_y$, and L$_z$ don't commute, you can't create an eigenstate that lists quantum numbers for any two of them. But because L$^2$ commutes with them, you can construct eigenstates that have eigenvalues for $L^2$ and any *one* of $L_x$, $L_y$, and $L_z$. By convention, the direction that's usually chosen is $L_z$.

# Creating the Angular Momentum Eigenstates

Now's the time to create the actual eigenstates, $|l, m>$, of angular momentum states in quantum mechanics. When you have the eigenstates, you'll also have the eigenvalues, and when you have the eigenvalues, you can solve the Hamiltonian and get the allowed energy levels of an object with angular momentum.

Don't make the assumption that the eigenstates are $|l, m>$; rather, say they're $|\alpha, \beta>$, where the eigenvalue of $L^2$ is $L^2|\alpha, \beta> = \hbar^2\alpha|\alpha, \beta>$. So the eigenvalue of $L^2$ is $\hbar^2\alpha$, where you have yet to solve for $\alpha$. Similarly, the eigenvalue of $L_z$ is $L_z|\alpha, \beta> = \hbar\beta|\alpha, \beta>$.

To proceed further, you have to introduce *raising* and *lowering* operators (as you do with the harmonic oscillator in Chapter 4). That way, you can solve for the ground state by, for example, applying the lowering operator to the ground state and setting the result equal to zero — and then solving for the ground state itself.

In this case, the raising operator is $L_+$ and the lowering operator is $L_-$. These operators raise and lower the $L_z$ quantum number. In a way analogous to the raising and lowering operators in Chapter 4, you can define the raising and lowering operators this way:

✔ **Raising:** $L_+ = L_x + iL_y$

✔ **Lowering:** $L_- = L_x - iL_y$

These two equations mean that

$$L_x = \frac{1}{2}\left(L_+ + L_-\right)$$

$$L_y = \frac{1}{2}\left(L_+ - L_-\right)$$

You can also see that

$$L_+L_- = L_x^2 + L_y^2 - \hbar L_z = L^2 - L_z^2 - \hbar L_z$$

That means the following are all equal to $L^2$:

- ✔ $L^2 = L_+ L_- + L_z^2 - \hbar L_-$
- ✔ $L^2 = L_- L_+ + L_z^2 + \hbar L_z$
- ✔ $L^2 = \frac{1}{2}(L_+ L_- + L_- L_+) + L_z^2$

You can also see that these equations are true:

- ✔ $[L^2, L_\pm] = 0$
- ✔ $[L_+, L_-] = 2\hbar L_z$
- ✔ $[L_z, L_\pm] = \pm\hbar L_\pm$

Okay, now you can put all this to work. You're getting to the good stuff.

Take a look at the operation of $L_+$ on $|\alpha, \beta\rangle$:

$$L_+ |\alpha, \beta\rangle = ?$$

To see what $L_+ |\alpha, \beta\rangle$ is, start by applying the $L_z$ operator on it like this:

$$L_z L_+ |\alpha, \beta\rangle = ?$$

From $[L_z, L_\pm] = \pm\hbar L_\pm$, you can see that $L_z L_+ - L_+ L_z = \hbar L_+$, so

$$L_z L_+ |\alpha, \beta\rangle = L_+ L_z |\alpha, \beta\rangle + \hbar L_+ |\alpha, \beta\rangle$$

And because $L_z |\alpha, \beta\rangle = \beta$, you have the following:

$$L_z L_+ |\alpha, \beta\rangle = \hbar(\beta + 1)L_+ |\alpha, \beta\rangle$$

This equation means that the eigenstate $L_+ |\alpha, \beta\rangle$ is also an eigenstate of the $L_z$ operator, with an eigenvalue of $(\beta + 1)$. Or in a more comprehensible way:

$$L_+ |\alpha, \beta\rangle = c |\alpha, \beta + 1\rangle$$

where $c$ is a constant you find later in "Finding the Eigenvalues of the Raising and Lowering Operators."

So the $L_+$ operator has the effect of rasing the $\beta$ quantum number by 1. Similarly, the lowering operator does this:

$$L_- |\alpha, \beta\rangle = c |\alpha, \beta - 1\rangle$$

Now take a look at what $L^2 L_+ |\alpha, \beta>$ equals:

$$L^2 L_+ |\alpha, \beta> = ?$$

Because $L^2$ is a scalar, it commutes with everything. $L^2 L_+ - L_+ L^2 = 0$, so this is true:

$$L^2 L_+ |\alpha, \beta> = L_+ L^2 |\alpha, \beta>$$

And because $L^2 |\alpha, \beta> = \alpha \hbar^2 |\alpha, \beta>$, you have the following equation:

$$L^2 L_+ |\alpha, \beta> = \alpha \hbar^2 L_+ |\alpha, \beta>$$

Similarly, the lowering operator, $L_-$, gives you this:

$$L^2 L_- |\alpha, \beta> = \alpha \hbar^2 L_- |\alpha, \beta>$$

So the results of these equations mean that the $L_\pm$ operators don't change the $\alpha$ eigenvalue of $|\alpha, \beta>$ at all.

Okay, so just what *are* $\alpha$ and $\beta$? Read on.

# Finding the Angular Momentum Eigenvalues

The eigenvalues of the angular momentum are the possible values the angular momentum can take, so they're worth finding. Let's take a look at how to do just that.

## Deriving eigenstate equations with $\beta_{max}$ and $\beta_{min}$

Note that $L^2 - L_z^2 = L_x^2 + L_y^2$, which is a positive number, so $L^2 - L_z^2 \geq 0$. That means that

$$<\alpha, \beta | L^2 - L_z^2 |\alpha, \beta> \geq 0$$

And substituting in $L^2 | \alpha, \beta> = \alpha \hbar^2$ and $L_z^2 | \alpha, \beta> = \beta \hbar$ gives you this:

$$<\alpha, \beta | L^2 - L_z^2 | \alpha, \beta> = \hbar^2(\alpha - \beta^2) \geq 0$$

Therefore, $\alpha \geq \beta^2$. So there's a maximum possible value of $\beta$, which you can call $\beta_{max}$.

You can be clever now, because there has to be a state $| \alpha, \beta_{max}>$ such that you can't raise $\beta$ any more. Thus, if you apply the raising operator, you get zero:

$$L_+ | \alpha, \beta_{max}> = 0$$

Applying the lowering operator to this also gives you zero:

$$L_- L_+ | \alpha, \beta_{max}> = 0$$

And because $L_- L_+ = L^2 - L_z^2 - \hbar L_z$, that means the following is true:

$$(L^2 - L_z^2 - \hbar L_z) | \alpha, \beta_{max}> = 0$$

Putting in $L^2 | \alpha, \beta_{max}> = \alpha \hbar^2$ and $L_z^2 | \alpha, \beta v> = \beta_{max} \hbar$ gives you this:

$$(\alpha - \beta_{max}^2 - \beta_{max}) \hbar^2 = 0$$
$$\alpha = \beta_{max}(\beta_{max} + 1) = 0$$

Cool, now you know what $\alpha$ is. At this point, it's usual to rename $\beta_{max}$ as $l$ and $\beta$ as $m$, so $| \alpha, \beta>$ becomes $| l, m>$ and

- ✔ $L^2 | l, m> = l(l + 1) \hbar^2 | l, m>$
- ✔ $L_z | l, m> = m \hbar | l, m>$

You can say even more. In addition to a $\beta_{max}$, there must also be a $\beta_{min}$ such that when you apply the lowering operator, $L_-$, you get zero, because you can't go any lower than $\beta_{min}$:

$$L_- | l, \beta_{min}> = 0$$

And you can apply $L_+$ on this as well:

$$L_+ L_- | l, \beta_{min}> = 0$$

From $L_-L_+ = L^2 - L_z^2 - \hbar L_z$, you know that

$$(L^2 - L_z^2 + \hbar L_z)|\alpha, \beta_{min}> = 0$$

which gives you the following:

$$(\alpha - \beta_{min}^2 + \beta_{min})\hbar^2 = 0$$
$$\alpha - \beta_{min}^2 + \beta_{min} = 0$$
$$\alpha = \beta_{min}^2 - \beta_{min}$$
$$\alpha = \beta_{min}(\beta_{min} - 1)$$

And comparing this equation to $\alpha = \beta_{max}(\beta_{max} + 1) = 0$ gives you

$$\beta_{max} = -\beta_{min}$$

Note that because you reach $|\alpha, \beta_{min}>$ by $n$ successive applications of $L_-$ on $|\alpha, \beta_{max}>$, you get the following:

$$\beta_{max} = \beta_{min} + n$$

Coupling these two equations gives you

$$\beta_{max} = {}^n/_2$$

Therefore, $\beta_{max}$ can be either an integer or half an integer (depending on whether $n$ is even or odd).

Because $l = \beta_{max}$, $m = \beta$, and $n$ is a positive number, you can find that $-l \le m \le l$. So now you have it:

- ✔ The eigenstates are $|l, m>$.
- ✔ The quantum number of the total angular momentum is $l$.
- ✔ The quantum number of the angular momentum along the $z$ axis is $m$.
- ✔ $L^2|l, m> = l(l + 1)|l, m>$, where $l = 0, {}^1/_2, 1, {}^3/_2, ...$
- ✔ $L_z|l, m> = m|l, m>$, where $m = -l, -(l - 1), ..., l - 1, l$.
- ✔ $-l \le m \le l$.

For each $l$, there are $2l + 1$ values of $m$. For example, if $l = 2$, then $m$ can equal $-2, -1, 0, 1,$ or $2$. If $l = {}^5/_2$, then $m$ can equal $-{}^5/_2, -{}^3/_2, -{}^1/_2, {}^1/_2, {}^3/_2,$ and ${}^5/_2$.

You can see a representative L and $L_z$ in Figure 5-2. L is the total angular momentum and $L_z$ is the projection of that total angular momentum on the z axis.

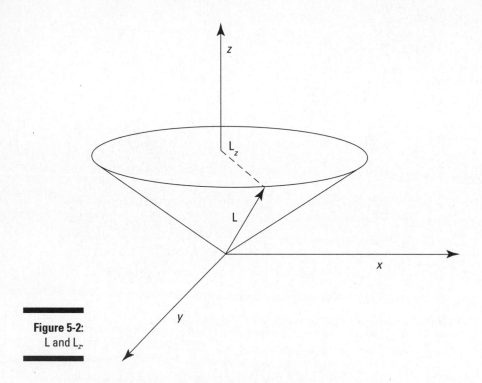

**Figure 5-2:**
L and $L_z$.

# Getting rotational energy of a diatomic molecule

Here's an example that involves finding the rotational energy spectrum of a diatomic molecule. Figure 5-3 shows the setup: A rotating diatomic molecule is composed of two atoms with masses $m_1$ and $m_2$. The first atom rotates at $r = r_1$, and the second atom rotates at $r = r_2$. What's the molecule's rotational energy?

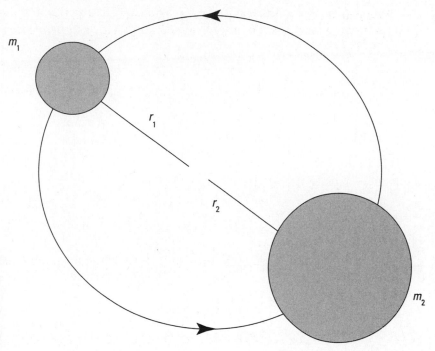

**Figure 5-3:**
A rotating
diatomic
molecule.

The Hamiltonian (as you can see at the chapter intro) is

$$H = \frac{L^2}{2I}$$

I is the rotational moment of inertia, which is

$$I = m_1 r_1^2 + m_2 r_2^2 = \mu r^2$$

where $r = |r_1 - r_2|$ and $\mu = \dfrac{m_1 m_2}{m_1 + m_2}$.

Because $L = I\omega$, $L = \mu r^2 \omega$. Therefore, the Hamiltonian becomes

$$H = \frac{L^2}{2I} = \frac{L^2}{2\mu r^2}$$

So applying the Hamiltonian to the eigenstates, $|l, m>$, gives you the following:

$$H|l,m> = \frac{L^2}{2\mu r^2}l,m$$

And as you know, $L^2|l, m> = l(l + 1)\hbar^2|l, m>$, so this equation becomes

$$H|l,m> = \frac{L^2}{2\mu r^2}l,m> = \frac{l(l+1)\hbar^2}{2\mu r^2}|l,m>$$

And because $H|l, m> = E|l, m>$, you can see that

$$E = \frac{l(l+1)\hbar^2}{2\mu r^2}$$

And that's the energy as a function of $l$, the angular momentum quantum number.

# Finding the Eigenvalues of the Raising and Lowering Operators

This section looks at finding the eigenvalues of the raising and lowering angular momentum operators, which raise and lower a state's $z$ component of angular momentum.

Start by taking a look at $L_+$, and plan to solve for $c$:

$$L_+|l, m> = c|l, m + 1>$$

So $L_+|l, m>$ gives you a new state, and multiplying that new state by its transpose should give you $c^2$:

$$(L_+|l, m>)^\dagger L_+|l, m> = c^2$$

To see this equation, note that $(L_+|l, m>)^\dagger L_+|l, m> = c^2<l, m + 1|l, m + 1> = c^2$. On the other hand, also note that $(L_+|l, m>)^\dagger L_+|l, m> = <l, m|L_+ L_-|l, m>$, so you have

$$<l, m|L_+ L_-|l, m> = c^2$$

What do you do about $L_+ L_-$? Well, you see earlier in the chapter, in "Creating the Angular Momentum Eigenstates," that this is true: $L_+ L_- = L^2 - L_z^2 + \hbar L_z$. So your equation becomes the following:

$$<l, m | L^2 - L_z^2 + \hbar L_z | l, m> = c^2$$

Great! That means that $c$ is equal to

$$c = (<l, m | L^2 - L_z^2 + \hbar L_z | l, m>)^{1/2}$$

So what is $(<l, m | L^2 - L_z^2 + \hbar L_z | l, m>)^{1/2}$? Applying the $L^2$ and $L_z$ operators gives you this value for $c$:

$$c = \hbar [l(l + 1) - m(m + 1)]^{1/2}$$

And that's the eigenvalue of $L_+$, which means you have this relation:

$$L_+ | l, m> = \hbar [l(l + 1) - m(m + 1)]^{1/2} | l, m + 1>$$

Similarly, you can show that $L_-$ gives you the following:

$$L_- | l, m> = \hbar [l(l + 1) - m(m - 1)]^{1/2} | l, m - 1>$$

# Interpreting Angular Momentum with Matrices

Chapter 4 covers a matrix interpretation of harmonic oscillator states and operators, and you can handle angular momentum the same way (which often makes understanding what's going on with angular momentum easier). You get to take a look at the matrix representation of angular momentum on a quantum level now.

Consider a system with angular momentum, with the total angular momentum quantum number $l = 1$. That means that $m$ can take the values –1, 0, and 1. So you can represent the three possible angular momentum states like this:

$$|1,-1> = \begin{vmatrix} 0 \\ 0 \\ 1 \end{vmatrix}$$

$|1,0>=$

$$\begin{vmatrix} 0 \\ 1 \\ 0 \end{vmatrix}$$

$|1,-1>=$

$$\begin{vmatrix} 1 \\ 0 \\ 0 \end{vmatrix}$$

Okay, so what are the operators you've seen in this chapter in matrix representation? For example, what is $L^2$? You can write $L^2$ this way in matrix form:

$L^2=$

$$\begin{vmatrix} <1,1|L^2|1,1> & <1,1|L^2|1,0> & <1,1|L^2|1,-1> \\ <1,0|L^2|1,1> & <1,0|L^2|1,0> & <1,0|L^2|1,-1> \\ <1,-1|L^2|1,1> & <1,-1|L^2|1,0> & <1,-1|L^2|1,-1> \end{vmatrix}$$

Okay, $<1, 1|L^2|1, 1> = l(l + 1)\hbar^2 = 2\hbar^2$; $<1, 1|L^2|1, 0> = 0$; $<1, 0|L^2|1, 0> = 2\hbar^2$; and so on; Therefore, the preceding matrix becomes the following:

$L^2 =$

$$\begin{vmatrix} 2\hbar^2 & 0 & 0 \\ 0 & 2\hbar^2 & 0 \\ 0 & 0 & 2\hbar^2 \end{vmatrix}$$

And you can also write this as

$L^2 =$

$$2\hbar^2 \begin{vmatrix} 1 & 0 & 0 \\ 0 & 1 & 0 \\ 0 & 0 & 1 \end{vmatrix}$$

So in matrix form, the equation $L^2|1, 1> = 2\hbar^2|1, 1>$ becomes

$$2\hbar^2 \begin{vmatrix} 1 & 0 & 0 \\ 0 & 1 & 0 \\ 0 & 0 & 1 \end{vmatrix} \begin{vmatrix} 1 \\ 0 \\ 0 \end{vmatrix}$$

This equals the following:

$$2\hbar^2 \begin{vmatrix} 1 \\ 0 \\ 0 \end{vmatrix} = 2\hbar^2 \begin{vmatrix} 1 & 0 & 0 \\ 0 & 1 & 0 \\ 0 & 0 & 1 \end{vmatrix} \begin{vmatrix} 1 \\ 0 \\ 0 \end{vmatrix}$$

How about the $L_+$ operator? As you probably know (from the preceding section), $L_+ | l, m> = \hbar[l(l + 1) - m(m + 1)]^{1/2} | l, m + 1>$. In this example, $l = 1$ and $m = 1, 0$, and $-1$. So you have the following:

- $L_+ | 1, 1> = 0$
- $L_+ |1,0> = \sqrt{2}\hbar|1,1>$
- $L_+ |1,-1> = \sqrt{2}\hbar|1,0>$

So the $L_+$ operator looks like this in matrix form:

$$L_+ =$$

$$\sqrt{2}\hbar \begin{vmatrix} 0 & 1 & 0 \\ 0 & 0 & 1 \\ 0 & 0 & 0 \end{vmatrix}$$

Therefore, $L_+ | 1, 0>$ would be

$$L_+ |1,0> =$$

$$\sqrt{2}\hbar \begin{vmatrix} 0 & 1 & 0 \\ 0 & 0 & 1 \\ 0 & 0 & 0 \end{vmatrix} \begin{vmatrix} 0 \\ 1 \\ 0 \end{vmatrix}$$

And this equals

$$L_+ |1,0> =$$

$$\sqrt{2}\hbar \begin{vmatrix} 1 \\ 0 \\ 0 \end{vmatrix} = \sqrt{2}\hbar \begin{vmatrix} 0 & 1 & 0 \\ 0 & 0 & 1 \\ 0 & 0 & 0 \end{vmatrix} \begin{vmatrix} 0 \\ 1 \\ 0 \end{vmatrix}$$

In other words, $\sqrt{2}\hbar|1,1> = L_+ |1,0>$ .

Okay, what about L_? You know that $L_-|l, m> = \hbar[l(l + 1) - m(m - 1)]^{1/2}|l, m - 1>$. In this example, $l = 1$ and $m = 1$, 0, and −1. So that means the following:

- $L_-|1,1> = \sqrt{2}\hbar|1,0>$
- $L_-|1,0> = \sqrt{2}\hbar|1,-1>$
- $L_-|1, -1> = 0$

So the L_ operator looks like this in matrix form:

$$L_- = \sqrt{2}\hbar\begin{vmatrix} 0 & 0 & 0 \\ 1 & 0 & 0 \\ 0 & 1 & 0 \end{vmatrix}$$

That means that $L_-|1, 1>$ would be

$$L_- |1,1> = \sqrt{2}\hbar\begin{vmatrix} 0 & 0 & 0 \\ 1 & 0 & 0 \\ 0 & 1 & 0 \end{vmatrix}\begin{vmatrix} 1 \\ 0 \\ 0 \end{vmatrix}$$

This equals

$$L_- |1,1> = \sqrt{2}\hbar\begin{vmatrix} 0 \\ 1 \\ 0 \end{vmatrix} = \sqrt{2}\hbar\begin{vmatrix} 0 & 0 & 0 \\ 1 & 0 & 0 \\ 0 & 1 & 0 \end{vmatrix}\begin{vmatrix} 1 \\ 0 \\ 0 \end{vmatrix}$$

Which tells you that

$$\sqrt{2}\hbar|1,0> = L_-|1,1>$$

Just as you'd expect.

Okay, you've found $L^2$, $L_+$, and $L_-$. Finding the matrix representation of $L_z$ is simple because

> ✔ $\hbar\,|1, 1> = L_z|1, 1>$
>
> ✔ $0 = L_z|1, 0>$
>
> ✔ $-\hbar\,|1, 1> = L_z|1, -1>$

So you have that

$$L_z = \hbar\begin{vmatrix} 1 & 0 & 0 \\ 0 & 0 & 0 \\ 0 & 0 & -1 \end{vmatrix}$$

Thus $L_z\,|1, -1>$ equals

$$L_z\left|1,-1\right> = \hbar\begin{vmatrix} 1 & 0 & 0 \\ 0 & 0 & 0 \\ 0 & 0 & -1 \end{vmatrix}\begin{vmatrix} 0 \\ 0 \\ -1 \end{vmatrix}$$

And this equals

$$L_z\left|1,-1\right> = -\hbar\begin{vmatrix} 0 \\ 0 \\ 1 \end{vmatrix} = \hbar\begin{vmatrix} 1 & 0 & 0 \\ 0 & 0 & 0 \\ 0 & 0 & -1 \end{vmatrix}\begin{vmatrix} 0 \\ 0 \\ -1 \end{vmatrix}$$

So $L_z|1, -1> = -\hbar\,|1, -1>$.

Now what about finding the $L_x$ and $L_y$ operators? That's not as hard as you may think, because

$$L_x = \frac{1}{2}\left(L_+ + L_-\right)$$

and

$$L_y = \frac{i}{2}\left(L_- - L_+\right)$$

Take a look at $L_x$ first. $L_+$ equals

$$L_+ = \sqrt{2}\hbar \begin{vmatrix} 0 & 1 & 0 \\ 0 & 0 & 1 \\ 0 & 0 & 0 \end{vmatrix}$$

And $L_-$ equals

$$L_- = \sqrt{2}\hbar \begin{vmatrix} 0 & 0 & 0 \\ 1 & 0 & 0 \\ 0 & 1 & 0 \end{vmatrix}$$

So this equals:

$$L_x = \frac{\hbar}{\sqrt{2}} \begin{vmatrix} 0 & 1 & 0 \\ 1 & 0 & 1 \\ 0 & 1 & 0 \end{vmatrix}$$

Okay, now what about $L_y$? $L_y = \dfrac{-i(L - L_1)}{2}$ , so:

$$L_y = \frac{\hbar}{\sqrt{2}} \begin{vmatrix} 0 & -i & 0 \\ i & 0 & -i \\ 0 & i & 0 \end{vmatrix}$$

Cool. This is going pretty well — how about calculating $[L_x, L_y]$? To do that, you need to calculate $[L_x, L_y] = L_xL_y - L_yL_x$. First find $L_xL_y$:

$$L_xL_y = \frac{\hbar^2}{2} \begin{vmatrix} 0 & 1 & 0 \\ 1 & 0 & 1 \\ 0 & 1 & 0 \end{vmatrix} \begin{vmatrix} 0 & -i & 0 \\ i & 0 & -i \\ 0 & i & 0 \end{vmatrix}$$

This equals

$$L_y L_x =$$

$$\frac{\hbar^2}{2} \begin{vmatrix} 0 & 1 & 0 \\ 1 & 0 & 1 \\ 0 & 1 & 0 \end{vmatrix} \begin{vmatrix} 0 & -i & 0 \\ i & 0 & -i \\ 0 & i & 0 \end{vmatrix} = \frac{\hbar^2}{2} \begin{vmatrix} i & 0 & -i \\ 1 & 0 & 0 \\ i & 0 & -i \end{vmatrix}$$

And similarly, $L_y L_x$ equals

$$L_y L_x =$$

$$\frac{\hbar^2}{2} \begin{vmatrix} 0 & -i & 0 \\ i & 0 & -i \\ 0 & i & 0 \end{vmatrix} \begin{vmatrix} 0 & 1 & 0 \\ 1 & 0 & 1 \\ 0 & 1 & 0 \end{vmatrix}$$

And this equals

$$L_y L_x =$$

$$\frac{\hbar^2}{2} \begin{vmatrix} 0 & -i & 0 \\ i & 0 & -i \\ 0 & i & 0 \end{vmatrix} \begin{vmatrix} 0 & 1 & 0 \\ 1 & 0 & 1 \\ 0 & 1 & 0 \end{vmatrix} = \begin{vmatrix} -i & 0 & -i \\ 0 & 0 & 0 \\ i & 0 & i \end{vmatrix}$$

So

$$\left[ L_x, L_y \right] = L_x L_y - L_y L_x =$$

$$\frac{\hbar^2}{2} \begin{vmatrix} 2i & 0 & 0 \\ 0 & 0 & 0 \\ 0 & 0 & -2i \end{vmatrix}$$

And this equals

$$\left[ L_x, L_y \right] = L_x L_y - L_y L_x =$$

$$\hbar^2 \begin{vmatrix} 1 & 0 & 0 \\ 0 & 0 & 0 \\ 0 & 0 & -1 \end{vmatrix}$$

But because

$$L_z = \hbar \begin{vmatrix} 1 & 0 & 0 \\ 0 & 0 & 0 \\ 0 & 0 & -1 \end{vmatrix}$$

you can rewrite $\left[ L_x, L_y \right] = L_x L_y - L_y L_x =$

$$\hbar^2 \begin{vmatrix} 1 & 0 & 0 \\ 0 & 0 & 0 \\ 0 & 0 & -1 \end{vmatrix}$$

like this: $\left[ L_x, L_y \right] = L_x L_y - L_y L_x = i\hbar L_z$

Cool, so $[L_x, L_y] = i\hbar L_z$.

# Rounding It Out: Switching to the Spherical Coordinate System

So far, this chapter has been dealing with angular momentum bras and kets:

$$\sqrt{2}\hbar \left| 1,0 \right> = L_- \left| 1,1 \right>$$

The charm of bras and kets is that they don't limit you to any specific system of representation (see Chapter 2). So you have the general eigenstates, but what are the actual *eigenfunctions* of $L_z$ and $L^2$? That is, you're going to try to find the actual functions that you can use with the angular momentum operators like $L^2$ and $L_z$.

To find the actual eigenfunctions (not just the eigenstates), you turn from rectangular coordinates, *x, y,* and *z,* to spherical coordinates because it'll make the math much simpler (after all, angular momentum is about things going around in circles). Figure 5-4 shows the spherical coordinate system.

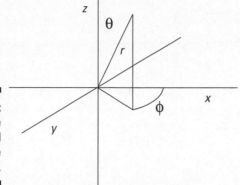

**Figure 5-4:**
The spherical coordinate system.

In the rectangular (Cartesian) coordinate system, you use $x$, $y$, and $z$ to orient yourself. In the spherical coordinate system, you also use three quantities: $r$, $\theta$, and $\phi$, as Figure 5-4 shows. You can translate between the spherical coordinate system and the rectangular one this way: The $r$ vector is the vector to the particle that has angular momentum, $\theta$ is the angle of $r$ from the z axis, and $\phi$ is the angle of $r$ from the x axis.

✔ $x = r\sin\theta\,\cos\phi$

✔ $y = r\sin\theta\,\sin\phi$

✔ $z = r\cos\theta$

Consider the equations for angular momentum:

$$L_x = YP_z - ZP_y = -i\hbar\left(\frac{Y\partial}{\partial z} - \frac{Z\partial}{\partial y}\right)$$

$$L_y = ZP_x - XP_z = -i\hbar\left(\frac{Z\partial}{\partial x} - \frac{X\partial}{\partial z}\right)$$

$$L_z = XP_y - YP_x = -i\hbar\left(\frac{X\partial}{\partial y} - \frac{Y\partial}{\partial x}\right)$$

When you take the angular momentum equations with the spherical-coordinate-system conversion equations, you can derive the following:

✔ $L^2 = \dfrac{-i\hbar\partial}{\partial\phi}$

✔ $L^2 = -\hbar^2\left(\dfrac{1}{\sin\theta}\dfrac{\partial}{\partial\theta}\left(\sin\theta\dfrac{\partial}{\partial\theta}\right) + \dfrac{1}{\sin^2\theta}\dfrac{\partial^2}{\partial\phi^2}\right)$

$$\text{✓ } L_+ = L_x + iL_y = \hbar e^{i\phi}\left(\frac{\partial}{\partial\theta} + \frac{i\cos\theta}{\sin\theta}\frac{\partial^2}{\partial\phi^2}\right)$$

$$\text{✓ } L_- = L_x - iL_y = -\hbar e^{-i\phi}\left(\frac{\partial}{\partial\theta} + \frac{i\cos\theta}{\sin\theta}\frac{\partial}{\partial\phi}\right)$$

Okay, these equations look pretty involved. But there's one thing to notice: They depend only on $\theta$ and $\phi$, which means their eigenstates depend only on $\theta$ and $\phi$, not on $r$. So the eigenfunctions of the operators in the preceding list can be denoted like this:

$<\theta, \phi | l, m>$

Traditionally, you give the name $Y_{lm}(\theta, \phi)$ to the eigenfunctions of angular momentum in spherical coordinates, so you have the following:

$Y_{lm}(\theta, \phi) = <\theta, \phi | l, m>$

All right, time to work on finding the actual form of $Y_{lm}(\theta, \phi)$. You know that when you use the $L^2$ and $L_z$ operators on angular momentum eigenstates, you get this:

$L^2 | l, m> = l(l + 1)\,\hbar^2 | l, m>$

$L_z | l, m> = m\hbar | l, m>$

So the following must be true:

$$\text{✓ } L^2 Y_{lm}(\theta, \phi) = l(l + 1)\hbar^2 Y_{lm}(\theta, \phi)$$

$$\text{✓ } L_z Y_{lm}(\theta, \phi) = m\hbar Y_{lm}(\theta,\phi)$$

In fact, you can go further. Note that $L_z$ depends only on $\theta$, which suggests that you can split $Y_{lm}(\theta,\phi)$ up into a part that depends on $\theta$ and a part that depends on $\phi$. Splitting $Y_{lm}(\theta,\phi)$ up into parts looks like this:

$Y_{lm}(\theta,\phi) = \Theta_{lm}(\theta)\Phi_m(\phi)$

That's what makes working with spherical coordinates so helpful — you can split the eigenfunctions up into two parts, one that depends only on $\theta$ and one part that depends only on $\phi$.

# The eigenfunctions of $L_z$ in spherical coordinates

Start by finding the eigenfunctions of $L_z$ in spherical coordinates. In spherical coordinates, the $L_z$ operator looks like this:

$$L_z = -i\hbar \frac{\partial}{\partial \phi}$$

So $L_z Y_{lm}(\theta, \phi) = L_z \Theta_{lm}(\theta)\Phi_m(\phi)$ is

$$L_z \Theta_{lm}(\theta)\Phi_m(\phi) = -i\hbar \frac{\partial}{\partial \phi}\Theta_{lm}(\theta)\Phi_m(\phi)$$

which is the following:

$$L_z \Theta_{lm}(\theta)\Phi_m(\phi) = -i\hbar \Theta_{lm}(\theta)\frac{\partial \Phi_m}{\partial \phi}(\phi)$$

And because $L_z Y_{lm}(\theta,\phi) = m\hbar Y_{lm}(\theta,\phi)$, this equation can be written in this version:

$$-i\hbar \Theta_{lm}(\theta)\frac{\partial \Phi_m}{\partial \phi}(\phi) = m\hbar \Theta_{lm}(\theta)\Phi_m(\phi)$$

Cancelling out terms from the two sides of this equation gives you this differential equation:

$$-i\frac{\partial \Phi_m}{\partial \phi}(\phi) = m\Phi_m(\phi)$$

This looks easy to solve, and the solution is just

$$\Phi_m(\phi) = Ce^{im\phi}$$

where C is a constant of integration.

You can determine C by insisting that $\Phi_m(\phi)$ be normalized — that is, that the following hold true:

$$\int_0^{2\pi} \Phi_m^*(\phi)\Phi_m(\phi)d\phi = 1$$

which gives you

$$C = \frac{1}{(2\pi)^{\frac{1}{2}}}$$

So $\Phi_m(\phi)$ is equal to this:

$$\Phi_m(\phi) = \frac{e^{im\phi}}{(2\pi)^{\frac{1}{2}}}$$

You're making progress — you've been able to determine the form of $\Phi_m(\phi)$, so $Y_{lm}(\theta,\phi) = \Theta_{lm}(\theta)\,\Phi_m(\phi)$, which equals

$$Y_{lm}(\theta,\phi) = \Theta_{lm}(\theta)\Phi_m(\phi) = \Theta_{lm}(\theta)\frac{e^{im\phi}}{(2\pi)^{\frac{1}{2}}}$$

That's great — you're halfway there, but you still have to determine the form of $\Theta_{lm}(\theta)$, the eigenfunction of $L^2$. That's coming up next.

## *The eigenfunctions of $L^2$ in spherical coordinates*

Now you're going to tackle the eigenfunction of $L^2$, $\Theta_{lm}(\theta)$. You already know that in spherical coordinates, the $L^2$ operator looks like this:

$$L^2 = -\hbar\left( \frac{1}{\sin\theta}\frac{\partial}{\partial\theta}\left( \sin\theta\frac{\partial}{\partial\theta} \right) + \frac{1}{\sin^2\theta}\frac{\partial^2}{\partial\phi^2} \right)$$

That's quite an operator. And you know that

$$Y_{lm}(\theta,\phi) = \Theta_{lm}(\theta)\frac{e^{im\phi}}{(2\pi)^{\frac{1}{2}}}$$

So applying the $L^2$ operator to $Y_{lm}(\theta,\phi)$ gives you the following:

$$L^2 Y_{lm}(\theta,\phi) = \frac{-\hbar^2}{(2\pi)^{\frac{1}{2}}} - \left[ \frac{1}{\sin\theta}\frac{\partial}{\partial\theta}\left( \sin\theta\frac{\partial}{\partial\theta} \right) + \frac{1}{\sin^2\theta\partial\phi^2} \right]\Theta_{lm}(\theta)e^{im\phi}$$

And because $L^2Y_{lm}(\theta,\phi) = l(l+1)\hbar^2 Y_{lm}(\theta, \Phi) = l(l+1)\hbar^2 \Theta lm(\theta)\Phi_m(\phi)$, this equation becomes

$$\frac{-\hbar^2}{(2\pi)^{\frac{1}{2}}} - \left[ \frac{1}{\sin\theta}\frac{\partial}{\partial\theta}\left(\sin\theta\frac{\partial}{\partial\theta}\right) + \frac{1}{\sin^2\theta\partial\phi^2} \right]\Theta_{lm}(\theta)e^{im\phi}$$

$$= l(l+1)\hbar^2\Theta_{lm}(\theta)\frac{e^{im\phi}}{(2\pi)^{\frac{1}{2}}}$$

Wow, what have you gotten in to? Cancelling terms and subtracting the right-hand side from the left finally gives you this differential equation:

$$\left[ \frac{1}{\sin\theta}\frac{\partial}{\partial\theta}\left(\sin\theta\frac{\partial}{\partial\theta}\right) + \frac{1}{\sin^2\theta}\frac{\partial^2}{\partial\phi^2} \right]\Theta_{lm}(\theta)e^{im\phi} + l(l+1)\Theta_{lm}(\theta)e^{im\phi} = 0$$

Combining terms and dividing by $e^{im\phi}$ gives you the following:

$$\frac{1}{\sin\theta}\frac{\partial}{\partial\theta}\left[\sin\theta\frac{\partial}{\partial\theta}\Theta_{lm}(\theta)\right] + \left[l(l+1) - \frac{m^2}{\sin^2\theta}\right]\Theta_{lm}(\theta) = 0$$

Holy cow! Isn't there someone who's tried to solve this kind of differential equation before? Yes, there is. This equation is a *Legendre differential equation*, and the solutions are well-known. (Whew!) In general, the solutions take this form:

$$\Theta_{lm}(\theta) = C_{lm}P_{lm}(\cos\theta)$$

where $P_{lm}(\cos\theta)$ is the *Legendre function*.

So what are the Legendre functions? You can start by separating out the *m* dependence, which works this way with the Legendre functions:

$$P_{lm}(x) = (1-x^2)^{\frac{|m|}{2}}\frac{d^{|m|}}{dx^{|m|}}P_l(x)$$

where $P_l(x)$ is called a *Legendre polynomial* and is given by the Rodrigues formula:

$$P_l(x) = \frac{1}{2^l l!}\frac{d^l}{dx^l}(x^2-1)$$

You can use this equation to derive the first few Legendre polynomials like this:

- $P_0(x) = 1$
- $P_1(x) = x$
- $P_2(x) = \frac{1}{2}(3x^2 - 1)$
- $P_3(x) = \frac{1}{2}(5x^3 - 3x)$
- $P_4(x) = \frac{1}{8}(35x^4 - 30x^2 + 3)$
- $P_5(x) = \frac{1}{8}(63x^5 - 70x^3 + 15x)$

and so on. That's what the first few $P_l(x)$ polynomials look like. So what do the associated Legendre functions, $P_{lm}(x)$ look like? You can also calculate them. You can start off with $P_{l0}(x)$, where $m = 0$. Those are easy because $P_{l0}(x) = P_l(x)$, so

- $P_{10}(x) = x$
- $P_{20}(x) = \frac{1}{2}(3x^2 - 1)$
- $P_{30}(x) = \frac{1}{2}(5x^3 - 3x)$

Also, you can find that

- $P_{11}(x) = (1 - x^2)^{1/2}$
- $P_{21}(x) = 3x(1 - x^2)^{1/2}$
- $P_{22}(x) = 3(1 - x^2)^{1/2}$
- $P_{31}(x) = \frac{3}{2}(5x^2 - 1)(1 - x^2)^{\frac{1}{2}}$
- $P_{32}(x) = 15x(1 - x^2)$
- $P_{33}(x) = 15x(1 - x^2)^{3/2}$

These equations give you an overview of what the $P_{lm}$ functions look like, which means you're almost done. As you may recall, $\Theta_{lm}(\theta)$, the $\theta$ part of $Y_{lm}(\theta, \phi)$, is related to the $P_{lm}$ functions like this:

$$\Theta_{lm}(\theta) = C_{lm}P_{lm}(\cos\theta)$$

And now you know what the $P_{lm}$ functions look like, but what do $C_{lm}$, the constants, look like? As soon as you have those, you'll have the complete angular momentum eigenfunctions, $Y_{lm}(\theta, \phi)$, because $Y_{lm}(\theta, \phi) = \Theta_{lm}(\theta)\Phi_m(\phi)$.

You can go about calculating the constants $C_{lm}$ the way you always calculate such constants of integration in quantum physics — you normalize the eigenfunctions to 1. For $Y_{lm}(\theta,\phi) = \Theta_{lm}(\theta)\Phi_m(\phi)$, that looks like this:

$$\int_0^{2\pi} d\phi \int_0^{\pi} Y_{lm}^*(0,\phi)Y_{lm}(0,\phi)\sin\theta \; d\theta = 1$$

Substitute the following three quantities in this equation:

- $Y_{lm}(\theta, \phi) = \Theta_{lm}(\theta) \; \Phi_m(\phi)$
- $\Phi_m(\phi) = \dfrac{e^{lm\phi}}{(2\pi)^{1/2}}$
- $\Theta_{lm}(\theta) = C_{lm}P_{lm}(\cos\theta)$

You get the following:

$$\frac{|C_{lm}|^2}{2\pi} \int_0^{2\pi} d\phi \int_0^{\pi} |P_{lm}^*(\cos\theta)|^2 \sin\theta \; d\theta = 1$$

The integral over $\phi$ gives $2\pi$, so this becomes

$$|C_{lm}|^2 \int_0^{\pi} |P_{lm}^*(\cos\theta)|^2 \sin\theta \; d\theta = 1$$

You can evaluate the integral to this:

$$|C_{lm}|^2 \frac{2}{2l+1}\frac{(l+m)!}{(l-m)!} = 1$$

So in other words:

$$C_{lm} = (-1)^m \left[ \frac{(2l+1)(l-m)!}{2(l+m)!} \right]^{1/2} \text{ where } m \geq 0$$

Which means that

$$\Theta_{lm}(\theta) = (-1)^m \left[ \frac{(2l+1)(l-m)!}{2(l+m)!} \right]^{1/2} P_{lm}(\cos\theta) \text{ where } m \geq 0$$

So $Y_{lm}(\theta, \phi) = \Theta_{lm}(\theta)\Phi_m(\phi)$, which is the angular momentum eigenfunction in spherical coordinates, is

$$Y_{lm}(\theta,\phi) = (-1)^m \left[ \frac{(2l+1)(l-m)!}{4\pi(l+m)!} \right]^{1/2} P_{lm}(\cos\theta)e^{im\phi} \text{ where } m \geq 0$$

The functions given by this equation are called the *normalized spherical harmonics*. Here are what the first few normalized spherical harmonics look like:

- $Y_{00}(\theta,\phi) = \dfrac{1}{(4\pi)^{1/2}}$

- $Y_{10}(\theta, \phi) = (3/4\pi)^{1/2} \cos\theta$

- $Y_{1\pm1}(\theta,\phi) = \mp\left(\dfrac{3}{8}\pi\right)^{1/2} e^{\pm i\phi}\sin\theta$

- $Y_{20}(\theta, \phi) = (5/16\pi)^{1/2}(3\cos^2\theta - 1)$

- $Y_{2\pm1}(\theta,\phi) = \mp\left(\dfrac{15}{8}\pi\right)^{1/2} e^{\pm i\phi}\sin\theta$

- $Y_{2\pm2}(\theta, \phi) = (15/32\pi)^{1/2} e^{\pm 2i\phi}\sin^2\theta$

In fact, you can use these relations to convert the spherical harmonics to rectangular coordinates:

- $\sin\theta\cos\phi = \dfrac{x}{r}$

- $\sin\theta\sin\phi = \dfrac{y}{r}$

- $\cos\theta = \dfrac{z}{r}$

Substituting these equations into

$$Y_{lm}(\theta,\phi) = (-1)^m \left[ \frac{(2l+1)(l-m)!}{4\pi(l+m)!} \right]^{1/2} P_{lm}(\cos\theta)e^{im\phi} \text{ where } m \geq 0 \text{ gives you the}$$

spherical harmonics in rectangular coordinates:

- $Y_{00}(x,y,z) = \dfrac{1}{(4\pi)^{1/2}}$

- $Y_{10}(x,y,z) = \left(\dfrac{3}{4\pi}\right)^{1/2}\dfrac{z}{r}$

$\blacktriangleright$ $Y_{1\pm1}(x,y,z) = \mp\left(\dfrac{3}{8\pi}\right)^{\frac{1}{2}}(x\pm iy)\Big/r$

$\blacktriangleright$ $Y_{20}(x,y,z) = \left(\dfrac{5}{16\pi}\right)^{\frac{1}{2}}(3z^2-r^2)\Big/r^2$

$\blacktriangleright$ $Y_{2\pm1}(x,y,z) = \mp\left(\dfrac{15}{8\pi}\right)^{\frac{1}{2}}z(x\pm iy)\Big/r^2$

$\blacktriangleright$ $Y_{2\pm2}(x,y,z) = \mp\left(\dfrac{15}{32\pi}\right)^{\frac{1}{2}}(x^2-y^2\pm 2ixy)\Big/r^2$

# Chapter 6

# Getting Dizzy with Spin

• • • • • • • • • • • • • • • • • • • • • • • • • • • • • • • • • • • • • • • • • • • • • • • • •

*In This Chapter*

▶ Discovering spin with the Stern-Gerlach experiment

▶ Looking at eigenstates and spin notation

▶ Understanding fermions and bosons

▶ Comparing the spin operators with angular momentum operators

▶ Working with spin $^1/_2$ and Pauli matrices

• • • • • • • • • • • • • • • • • • • • • • • • • • • • • • • • • • • • • • • • • • • • • • • • •

*P*hysicists have suggested that orbital angular momentum is not the only kind of angular momentum present in an atom — electrons could also have *intrinsic* built-in angular momentum. This kind of built-in angular momentum is called *spin*. Whether or not electrons actually spin will never be known — they're as close to point-like particles as you can come, without any apparent internal structure. Yet the fact remains that they have intrinsic angular momentum. And that's what this chapter is about — the intrinsic, built-in quantum mechanical spin of subatomic particles.

# The Stern-Gerlach Experiment and the Case of the Missing Spot

The Stern-Gerlach experiment unexpectedly revealed the existence of spin back in 1922. Physicists Otto Stern and Walther Gerlach sent a beam of silver atoms through the poles of a magnet — whose magnetic field was in the $z$ direction — as you can see in Figure 6-1.

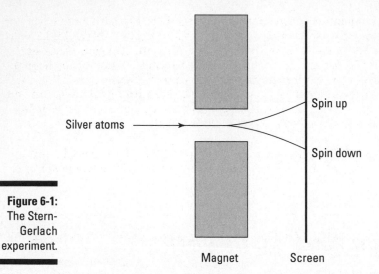

**Figure 6-1:**
The Stern-
Gerlach
experiment.

Because 46 of silver's 47 electrons are arranged in a symmetrical cloud, they contribute nothing to the orbital angular momentum of the atom. The 47th electron can be in

- The 5s state, in which case its angular momentum is $l = 0$ and the $z$ component of that angular momentum is 0

- The 5p state, in which case its angular momentum is $l = 1$, which means that the $z$ component of its angular momentum can be −1, 0, or 1

That means that Stern and Gerlach expected to see one or three spots on the screen you see at right in Figure 6-1, corresponding to the different states of the $z$ component of angular momentum.

But famously, they saw only two spots. This puzzled the physics community for about three years. Then, in 1925, physicists Samuel A. Goudsmit and George E. Uhlenbeck suggested that electrons contained intrinsic angular momentum — and that intrinsic angular momentum is what gave them a magnetic moment that interacted with the magnetic field. After all, it was apparent that some angular momentum other than orbital angular momentum was at work here. And that built-in angular momentum came to be called *spin*.

The beam of silver atoms divides in two, depending on the spin of the 47th electron in the atom, so there are two possible states of spin, which came to be known as *up* and *down*.

Spin is a purely quantum mechanical effect, and there's no real classical analog. The closest you can come is to liken spin to the spin of the Earth as it goes around the sun — that is, the Earth has both spin (because it's rotating on its axis) and orbital angular momentum (because it's revolving around the sun). But even this picture doesn't wholly explain spin in classical terms, because it's conceivable that you could stop the Earth from spinning. But you can't stop electrons from possessing spin, and that also goes for other subatomic particles that possess spin, such as protons.

Spin doesn't depend on spatial degrees of freedom; even if you were to have an electron at rest (which violates the uncertainty principle), it would still possess spin.

# Getting Down and Dirty with Spin and Eigenstates

Spin throws a bit of a curve at you. When dealing with orbital angular momentum (see Chapter 5), you can build angular momentum operators because orbital angular momentum is the product of momentum and radius. But spin is built in; there's no momentum operator involved. So here's the crux: You cannot describe spin with a differential operator, as you can for orbital angular momentum.

In Chapter 5, I show how orbital angular momentum can be reduced to these differential operators:

- $L_x = YP_z - ZP_y = -i\hbar\left(Y\dfrac{\partial}{\partial z} - Z\dfrac{\partial}{\partial y}\right)$

- $L_y = ZP_x - XP_z = -i\hbar\left(Z\dfrac{\partial}{\partial x} - X\dfrac{\partial}{\partial z}\right)$

- $L_z = XP_y - YP_x = -i\hbar\left(X\dfrac{\partial}{\partial y} - Y\dfrac{\partial}{\partial x}\right)$

And you can find eigenfunctions for angular momentum, such as $Y_{20}$:

$$Y_{20}(x,y,z) = \left(\frac{5}{16\pi}\right)^{1/2}\left(\frac{3z^2 - r^2}{r^2}\right)$$

But because you can't express spin using differential operators, you can't find eigenfunctions for spin as you do for angular momentum. So that means that you're left with the bra and ket way of looking at things (bras and kets aren't tied to any specific representation in spatial terms).

In Chapter 5, you also take a look at things in angular momentum terms, introducing the eigenstates of orbital angular momentum like this: $|l, m>$ (where $l$ is the angular momentum quantum number and $m$ is the quantum number of the $z$ component of angular momentum).

You can use the same notation for spin eigenstates. As with orbital angular momentum, you can use a total spin quantum number and a quantum number that indicates the spin along the $z$ axis (*Note:* There's no true $z$ axis built in when it comes to spin — you introduce a $z$ axis when you apply a magnetic field; by convention, the $z$ axis is taken to be in the direction of the applied magnetic field).

The letters given to the total spin quantum number and the $z$-axis component of the spin are $s$ and $m$ (you sometimes see them written as $s$ and $m_s$). In other words, the eigenstates of spin are written as $|s, m>$.

So what possible values can $s$ and $m$ take? That's coming up next.

# Halves and Integers: Saying Hello to Fermions and Bosons

In analogy with orbital angular momentum, you can assume that $m$ (the $z$-axis component of the spin) can take the values $-s$, $-s + 1$, ..., $s - 1$, and $s$, where $s$ is the total spin quantum number. For electrons, Stern and Gerlach observed two spots, so you have $2s + 1 = 2$, which means that $s = \frac{1}{2}$. And therefore, $m$ can be $+\frac{1}{2}$ or $-\frac{1}{2}$. So here are the possible eigenstates for electrons in terms of spin:

$$|\tfrac{1}{2}, \tfrac{1}{2}>$$

$$|\tfrac{1}{2}, -\tfrac{1}{2}>$$

So do all subatomic particles have $s = \frac{1}{2}$? Nope. Here are their options:

✔ **Fermions:** In physics, particles with half-integer spin are called *fermions*. They include electrons, protons, neutrons, and so on, even quarks. For example, electrons, protons, and neutrons have spin $s = \frac{1}{2}$, and delta particles have $s = \frac{3}{2}$.

✓ **Bosons:** Particles with integer spin are called *bosons.* They include photons, pi mesons, and so on; even the postulated particles involved with
the force of gravity, *gravitons,* are supposed to have integer spin. For
example, pi mesons have spin $s = 0$, photons have $s = 1$, and so forth.

So for electrons, the spin eigenstates are $|\frac{1}{2}, \frac{1}{2}>$ and $|\frac{1}{2}, -\frac{1}{2}>$. For photons,
the eigenstates are $|1, 1>$, $|1, 0>$, and $|1, -1>$. Therefore, the possible eigen-
states depend on the particle you're working with.

# Spin Operators: Running Around with Angular Momentum

Because spin is a type of built-in angular momentum, the spin operators have
a lot in common with the orbital angular momentum operators. In Chapter 5,
I discuss the orbital angular momentum operators $L^2$ and $L_z$, and as you may
expect, there are analogous spin operators, $S^2$ and $S_z$. However, these operators are just operators; they don't have a differential form like the orbital
angular momentum operators do.

In fact, all the orbital angular momentum operators, such as $L_x$, $L_y$, and $L_z$,
have analogs here: $S_x$, $S_y$, and $S_z$. The commutation relations among $L_x$, $L_y$,
and $L_z$ are the following:

✓ $[L_x, L_y] = i\hbar L_z$

✓ $[L_y, L_z] = i\hbar L_x$

✓ $[L_z, L_x] = i\hbar L_y$

And they work the same way for spin:

✓ $[S_x, S_y] = i\hbar S_z$

✓ $[S_y, S_z] = i\hbar S_x$

✓ $[S_z, S_x] = i\hbar S_y$

The $L^2$ operator gives you the following result when you apply it to an orbital
angular momentum eigenstate:

$L^2|l, m> = l(l + 1)\hbar^2|l, m>$

And just as you'd expect, the $S^2$ operator works in an analogous fashion:

$$S^2|s, m> = s(s + 1)\hbar^2|s, m>$$

The $L_z$ operator gives you this result when you apply it to an orbital angular momentum eigenstate (see Chapter 5):

$$L_z|l, m> = m\hbar|l, m>$$

And by analogy, the $S_z$ operator works this way:

$$S_z|s, m> = m\hbar|s, m>$$

What about the raising and lowering operators, $L_+$ and $L_-$? Are there analogs for spin? In angular momentum terms, $L_+$ and $L_-$ work like this:

- $L_+|l,m> = \hbar\left[l(l+1) - m(m+1)\right]^{\frac{1}{2}}|l,m+1>$

- $L_-|l,m> = \hbar\left[l(l+1) - m(m-1)\right]^{\frac{1}{2}}|l,m-1>$

There are spin raising and lowering operators as well, $S_+$ and $S_-$, and they work like this:

- $S_+|s,m> = \hbar\left[s(s+1) - m(m+1)\right]^{\frac{1}{2}}|s,m+1>$

- $S_-|s,m> = \hbar\left[s(s+1) - m(m-1)\right]^{\frac{1}{2}}|s,m-1>$

In the next section, I take a special look at particles with spin $\frac{1}{2}$.

# *Working with Spin ¹/₂ and Pauli Matrices*

Spin $\frac{1}{2}$ particles (fermions) need a little extra attention. The eigenvalues of the $S^2$ operator here are

$$S^2|s,m> = s(s+1)\hbar^2|s,m> = \frac{3}{4}\hbar^2|s,m>$$

And the eigenvalues of the $S_z$ operator are

$$S_z |s,m> \; = m\hbar |s,m> \; = \pm \frac{\hbar}{2} |s,m>$$

You can represent these two equations graphically as shown in Figure 6-2, where the two spin states have different projections along the *z* axis.

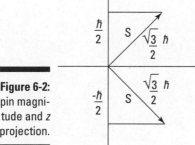

**Figure 6-2:** Spin magnitude and *z* projection.

## Spin ¹/₂ matrices

Time to take a look at the spin eigenstates and operators for particles of spin ¹/₂ in terms of matrices. There are only two possible states, spin up and spin down, so this is easy. First, you can represent the eigenstate |¹/₂, ¹/₂> like this:

$$\left| \frac{1}{2}, \frac{1}{2} \right> = \begin{vmatrix} 1 \\ 0 \end{vmatrix}$$

And the eigenstate |¹/₂, −¹/₂> looks like this:

$$\left| \frac{1}{2}, -\frac{1}{2} \right> = \begin{vmatrix} 0 \\ 1 \end{vmatrix}$$

Now what about spin operators like $S^2$? The $S^2$ operator looks like this in matrix terms:

$$S^2 =$$

$$\begin{vmatrix} \langle \tfrac{1}{2},\tfrac{1}{2}|S^2|\tfrac{1}{2},\tfrac{1}{2}\rangle & \langle \tfrac{1}{2},\tfrac{1}{2}|S^2|\tfrac{1}{2},-\tfrac{1}{2}\rangle \\ \langle \tfrac{1}{2},-\tfrac{1}{2}|S^2|\tfrac{1}{2},\tfrac{1}{2}\rangle & \langle \tfrac{1}{2},-\tfrac{1}{2}|S^2|\tfrac{1}{2},-\tfrac{1}{2}\rangle \end{vmatrix}$$

And this works out to be the following:

$$S^2 = \frac{3}{4}\hbar^2 \begin{vmatrix} 1 & 0 \\ 0 & 1 \end{vmatrix}$$

Similarly, you can represent the $S_z$ operator this way:

$$S_z =$$

$$\begin{vmatrix} \langle \tfrac{1}{2},\tfrac{1}{2}|S^2|\tfrac{1}{2},\tfrac{1}{2}\rangle & \langle \tfrac{1}{2},\tfrac{1}{2}|S^2|\tfrac{1}{2},-\tfrac{1}{2}\rangle \\ \langle \tfrac{1}{2},-\tfrac{1}{2}|S^2|\tfrac{1}{2},\tfrac{1}{2}\rangle & \langle \tfrac{1}{2},-\tfrac{1}{2}|S^2|\tfrac{1}{2},-\tfrac{1}{2}\rangle \end{vmatrix}$$

This works out to

$$S_z = \frac{\hbar}{2} \begin{vmatrix} 1 & 0 \\ 0 & -1 \end{vmatrix}$$

Using the matrix version of $S_z$, for example, you can find the $z$ component of the spin of, say, the eigenstate $|1/2, -1/2\rangle$. Finding the $z$ component looks like this:

$$S_z \; |1/2, -1/2\rangle$$

Putting this in matrix terms gives you this matrix product:

$$\frac{\hbar}{2} \begin{vmatrix} 1 & 0 \\ 0 & -1 \end{vmatrix} \begin{vmatrix} 0 \\ 1 \end{vmatrix}$$

Here's what you get by performing the matrix multiplication:

$$\frac{\hbar}{2} \begin{vmatrix} 1 & 0 \\ 0 & -1 \end{vmatrix} \begin{vmatrix} 0 \\ 1 \end{vmatrix} = \frac{-\hbar}{2} \begin{vmatrix} 0 \\ 1 \end{vmatrix}$$

And putting this back into ket notation, you get the following:

$$S_z \left| \frac{1}{2}, -\frac{1}{2} \right> = \frac{-\hbar}{2} \left| \frac{1}{2}, -\frac{1}{2} \right>$$

How about the raising and lowering operators $S_+$ and $S_-$? The $S_+$ operator looks like this:

$$S_+ =$$
$$\hbar \begin{vmatrix} 0 & 1 \\ 0 & 0 \end{vmatrix}$$

And the lowering operator looks like this:

$$S_- =$$
$$\hbar \begin{vmatrix} 0 & 0 \\ 1 & 0 \end{vmatrix}$$

So, for example, you can figure out what $S_+ | 1/2, -1/2>$ is. Here it is in matrix terms:

$$\hbar \begin{vmatrix} 0 & 1 \\ 0 & 0 \end{vmatrix} \begin{vmatrix} 0 \\ 1 \end{vmatrix}$$

Performing the multiplication gives you this:

$$\hbar \begin{vmatrix} 0 & 1 \\ 0 & 0 \end{vmatrix} \begin{vmatrix} 0 \\ 1 \end{vmatrix} = \hbar \begin{vmatrix} 1 \\ 0 \end{vmatrix}$$

Or in ket form, it's $S_+ | 1/2, -1/2> = \hbar | 1/2, 1/2>$. Cool.

## Pauli matrices

Sometimes, you see the operators $S_x$, $S_y$, and $S_z$ written in terms of *Pauli matrices*, $\sigma_x$, $\sigma_y$, and $\sigma_z$. Here's what the Pauli matrices look like:

$$\sigma_x =$$
$$\begin{vmatrix} 0 & 1 \\ 1 & 0 \end{vmatrix}$$

$$\sigma_y =$$
$$\begin{vmatrix} 0 & -i \\ i & 0 \end{vmatrix}$$

$$\sigma_z =$$
$$\begin{vmatrix} 1 & 0 \\ 0 & -1 \end{vmatrix}$$

Now you can write $S_x$, $S_y$, and $S_z$ in terms of the Pauli matrices like this:

$$S_x = \frac{\hbar}{2}\sigma_x$$

$$S_y = \frac{\hbar}{2}\sigma_y$$

$$S_z = \frac{\hbar}{2}\sigma_z$$

Whoo! And that concludes your look at spin.

# Part IV
# Multiple Dimensions: Going 3D with Quantum Physics

# In this part . . .

The previous parts deal mostly with particles in one-dimensional systems. This part expands that coverage to three dimensions, like in the real world. You see how to handle quantum physics in three-dimensional coordinates — whether rectangular or spherical — which lays the groundwork for working with electrons in atoms.

# Chapter 7

# Rectangular Coordinates: Solving Problems in Three Dimensions

- - - - - - - - - - - - - - - - - - - - - - - - - - - - - - - - - - - - - - - - - - - - -

### In This Chapter

▶ Exploring the Schrödinger equation in the $x$, $y$, and $z$ dimensions

▶ Working with free particles in 3D

▶ Getting into rectangular potentials

▶ Seeing harmonic oscillators in 3D space

- - - - - - - - - - - - - - - - - - - - - - - - - - - - - - - - - - - - - - - - - - - - -

*O*ne-dimensional problems are all very well and good, but the real world has three dimensions. This chapter is all about leaving one-dimensional potentials behind and starting to take a look at spinless quantum mechanical particles in three dimensions.

Here, you work with three dimensions in rectangular coordinates, starting with a look at the Schrödinger equation in glorious, real-life 3D. You then delve into free particles, box potentials, and harmonic oscillators. *Note:* By the way, the next chapter uses spherical coordinates because some problems are better in one system than the other. Problems with spherical symmetry are best handled in spherical coordinates, for example.

## The Schrödinger Equation: Now in 3D!

In one dimension, the time-dependent Schrödinger equation (of the type in Chapters 3 and 4 that let you find the wave function) looks like this:

$$\frac{-\hbar^2}{2m}\frac{\partial^2}{\partial x^2}\psi(x,t)+V(x,t)\psi(x,t)=i\hbar\frac{\partial}{\partial t}\psi(x,t)$$

And you can generalize that into three dimensions like this:

$$\frac{-\hbar^2}{2m}\left(\frac{\partial^2}{\partial x^2}+\frac{\partial^2}{\partial y^2}+\frac{\partial^2}{\partial z^2}\right)\psi(x,y,z,t)+V(r)\psi(x,y,z,t)=i\hbar\frac{\partial}{\partial t}\psi(x,y,z,t)$$

Using the Laplacian operator, you can recast this into a more compact form. Here's what the Laplacian looks like:

$$\left(\frac{\partial^2}{\partial x^2}+\frac{\partial^2}{\partial y^2}+\frac{\partial^2}{\partial z^2}\right)=\nabla^2$$

And here's the 3D Schrödinger equation using the Laplacian:

$$\frac{-\hbar^2}{2m}\nabla^2\psi(x,y,z,t)+V(x,y,z,t)\psi(x,y,z,t)=i\hbar\frac{\partial}{\partial t}\psi(x,y,z,t)$$

To solve this equation, break out the time-dependent part of the wave function:

$$\psi(x,y,z,t)=\psi(x,y,z)e^{-iEt/\hbar}$$

Here, $\psi(x, y, z)$ is the solution of the time-independent Schrödinger equation, and E is the energy:

$$\frac{-\hbar^2}{2m}\nabla^2\psi(x,y,z)+V(x,y,z)\psi(x,y,z)=E\psi(x,y,z)$$

So far, so good. But now you've run into a wall — the expression $\nabla^2\psi(x,y,z)$ is in general very hard to deal with, so the current equation is in general very hard to solve.

So what should you do? Well, you can focus on the case in which the equation is *separable* — that is, where you can separate out the *x, y,* and *z* dependence and find the solution in each dimension separately. In other words, in separable cases, the potential, V(*x, y, z*), is actually the sum of the *x, y,* and *z* potentials:

V(*x, y, z*) = V$_x$(*x*) + V$_y$(*y*) + V$_z$(*z*)

Now you can break the Hamiltonian in $\frac{-\hbar^2}{2m}\nabla^2\psi(x,y,z)+V(x,y,z)\psi(x,y,z)$ $=E\psi(x,y,z)$ into three Hamiltonians, H$_x$, H$_y$, and H$_z$:

$$(H_x + H_y + H_z)\psi(x, y, z) = E\psi(x, y, z)$$

where

$$\vdash H_x = \frac{-\hbar^2}{2m}\frac{\partial^2}{\partial x^2} + V_x(x)$$

$$\vdash H_y = \frac{-\hbar^2}{2m}\frac{\partial^2}{\partial y^2} + V_y(y)$$

$$\vdash H_z = \frac{-\hbar^2}{2m}\frac{\partial^2}{\partial z^2} + V_z(z)$$

When you divide up the Hamiltonian as in $(H_x + H_y + H_z)\psi(x, y, z) = E\psi(x, y, z)$, you can also divide up the wave function that solves that equation. In particular, you can break the wave function into three parts, one for $x$, $y$, and $z$:

$$\psi(x, y, z) = X(x)Y(y)Z(z)$$

That's going to make life considerably easier, because now you can break the Hamiltonian up into three separate operators added together:

$$\left(\frac{-\hbar^2}{2m}\frac{\partial^2}{\partial x^2} + V_x(x)\right)$$
$$+\left(\frac{-\hbar^2}{2m}\frac{\partial^2}{\partial y^2} + V_y(y)\right)$$
$$+\left(\frac{-\hbar^2}{2m}\frac{\partial^2}{\partial z^2} + V_z(z)\right)$$
$$= E$$

The total energy, E, is now the sum of the $x$ component's energy plus the $y$ component's energy plus the $z$ component's energy:

$$E = E_x + E_y + E_z$$

So you now have three independent Schrödinger equations for the three dimensions:

$$\vdash \frac{-\hbar^2}{2m}\frac{\partial^2}{\partial x^2}X(x) + V(x)X(x) = E_x X(x)$$

$$\vdash \frac{-\hbar^2}{2m}\frac{\partial^2}{\partial y^2}Y(y) + V(y)Y(y) = E_y Y(y)$$

$$\vdash \frac{-\hbar^2}{2m}\frac{\partial^2}{\partial z^2}Z(z) + V(y)Z(z) = E_z Z(z)$$

This system of independent differential equations looks a lot easier to solve than $(H_x + H_y + H_z)\psi(x, y, z) = E\psi(x, y, z)$. In essence, you've broken the three-dimensional Schrödinger equation into three one-dimensional Schrödinger equations. That makes solving 3D problems tractable.

# Solving Three-Dimensional Free Particle Problems

Consider the free particle you see in three dimensions in Figure 7-1.

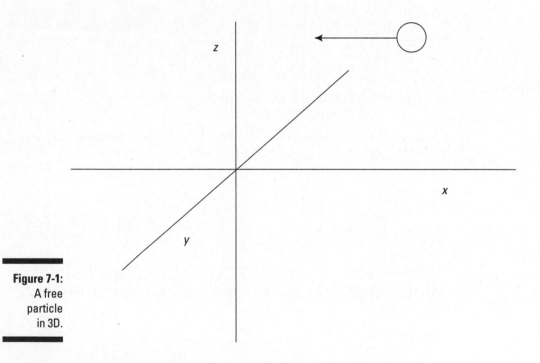

**Figure 7-1:**
A free
particle
in 3D.

Because the particle is traveling freely, $V(x) = V(y) = V(z) = 0$. So the three independent Schrödinger equations for the three dimensions covered in the preceding section become the following:

$$\frac{-\hbar^2}{2m}\frac{\partial^2}{\partial x^2}X(x) = E_x X(x)$$

$$\frac{-\hbar^2}{2m}\frac{\partial^2}{\partial y^2}Y(y) = E_y Y(y)$$

$$\frac{-\hbar^2}{2m}\frac{\partial^2}{\partial z^2}Z(z) = E_z Z(z)$$

If you rewrite these equations in terms of the wave number, $k$, where $k^2 = \frac{2mE}{\hbar^2}$, then these equations become the following:

$$\frac{\partial^2}{\partial x^2}X(x) = -k_x^2 X(x)$$

$$\frac{\partial^2}{\partial y^2}Y(y) = -k_y^2 Y(y)$$

$$\frac{\partial^2}{\partial z^2}Z(z) = -k_z^2 Z(z)$$

In this section, you take a look at the solutions to these equations, find the total energy, and add time dependence.

## The x, y, and z equations

Take a look at the $x$ equation for the free particle, $\frac{\partial^2}{\partial x^2}X(x) = -k_x^2 X(x)$. You can write its general solution as

$$X(x) = Ae^{ikx}$$

This is a plane wave, and normalizing it (as I discuss in Chapter 3) gives you this:

$$X(x) = \frac{1}{(2\pi)^{1/2}}e^{ik_x x}$$

The $y$ and $z$ components follow the same form:

$$Y(y) = \frac{1}{(2\pi)^{\frac{1}{2}}} e^{ik_y y}$$

$$Z(z) = \frac{1}{(2\pi)^{\frac{1}{2}}} e^{ik_z z}$$

Because $\psi(x, y, z) = X(x)Y(y)Z(z)$, you get this for $\psi(x, y, z)$:

$$\psi(x,y,z) = \frac{1}{(2\pi)^{\frac{3}{2}}} e^{ik_x x} e^{ik_y y} e^{ik_z z}$$

$$= \frac{1}{(2\pi)^{\frac{3}{2}}} e^{i(k_x x + k_y y + k_z z)}$$

The part in the parentheses in the exponent is the dot product of the vectors $\boldsymbol{k}$ and $\boldsymbol{r}$, $\boldsymbol{k} \cdot \boldsymbol{r}$. That is, if the vector $\boldsymbol{a} = (a_x, a_y, a_z)$ in terms of components and the vector $\boldsymbol{b} = (b_x, b_y, b_z)$, then the dot product of $\boldsymbol{a}$ and $\boldsymbol{b}$ is $\boldsymbol{a} \cdot \boldsymbol{b} = (a_x b_x, a_y b_y, a_z b_z)$. So here's how you can rewrite the $\psi(x, y, z)$ equation:

$$\psi(x,y,z) = \frac{1}{(2m)^{\frac{3}{2}}} e^{ik \cdot r}$$

## Finding the total energy equation

The total energy of the free particle is the sum of the energy in three dimensions:

$$E = E_x + E_y + E_z$$

With a free particle, the energy of the $x$ component of the wave function is $\frac{\hbar^2 k_x^{\ 2}}{2m} = E_x$. And this equation works the same way for the $y$ and $z$ components, so here's the total energy of the particle:

$$E = \frac{\hbar^2 k_x^{\ 2}}{2m} + \frac{\hbar^2 k_y^{\ 2}}{2m} + \frac{\hbar^2 k_z^{\ 2}}{2m}$$

$$= \frac{\hbar^2}{2m}\left(k_x^{\ 2} + k_y^{\ 2} + k_z^{\ 2}\right)$$

Note that $k_x{}^2 + k_y{}^2 + k_z{}^2$ is the square of the magnitude of $\boldsymbol{k}$ — that is, $\boldsymbol{k}^2$. Therefore, you can write the equation for the total energy as

$$E = \frac{\hbar^2}{2m}\left(k_x{}^2 + k_y{}^2 + k_z{}^2\right) = \frac{\hbar^2}{2m}\boldsymbol{k}^2$$

Note that because E is a constant, no matter where the particle is pointed, all the eigenfunctions of $\frac{\partial^2}{\partial x^2}X(x) = -k_x{}^2 X(x)$, $\frac{\partial^2}{\partial y^2}Y(y) = -k_y{}^2 Y(y)$, and $\frac{\partial^2}{\partial z^2}Z(z) = -k_z{}^2 Z(z)$ are infinitely degenerate as you vary $k_x$, $k_y$, and $k_z$.

## Adding time dependence and getting a physical solution

You can add time dependence to the solution for $\psi(x, y, z)$, giving you $\psi(x, y, z, t)$, if you remember that $\psi(x,y,z,t) = \psi(x,y,z)e^{-iEt/\hbar}$. That equation gives you this form for $\psi(x, y, z, t)$:

$$\psi(x,y,z,t) = \frac{1}{(2\pi)^{3/2}} e^{i\left(\boldsymbol{k}\cdot\boldsymbol{r} - Et/\hbar\right)}$$

Because $\omega = \dfrac{E}{\hbar}$, the equation turns into

$$\psi(x,y,z,t) = \frac{1}{(2\pi)^{3/2}} e^{i\left(\boldsymbol{k}\cdot\boldsymbol{r} - \omega t\right)}$$

In fact, now that the right side of the equation is in terms of the radius vector **r**, you can make the left side match:

$$\psi(\boldsymbol{r},t) = \frac{1}{(2\pi)^{3/2}} e^{i\left(\boldsymbol{k}\cdot\boldsymbol{r} - \omega t\right)}$$

That's the solution to the Schrödinger equation, but it's unphysical (as I discuss for the one-dimensional Schrödinger equation for a free particle in Chapter 3). Why? Trying to normalize this equation in three dimensions, for example, gives you the following, where C is a constant:

$$\int\limits_{-\infty}^{+\infty} \psi(r,t)\psi^*(r,t)d^3r$$

$$= |C|^2 \int\limits_{-\infty}^{+\infty} d^3r \rightarrow \infty$$

Thus, the integral diverges and you can't normalize $\psi(r,t)$ as I've written it. So what do you do here to get a physical particle?

The key to solving this problem is realizing that if you have a number of solutions to the Schrödinger equation, then any linear combination of those solutions is also a solution. In other words, you add various wave functions together so that you get a *wave packet,* which is a collection of wave functions of the form $e^{ik \cdot r}$ such that

  ✔ The wave functions interfere constructively at one location.

  ✔ They interfere destructively (go to zero) at all other locations.

Look at the time-independent version:

$$\psi(r) = \sum_{n=1}^{\infty} \phi_n e^{ik \cdot r}$$

However, for a free particle, the energy states are not separated into distinct bands; the possible energies are continuous, so people write this summation as an integral:

$$\psi(r) = \frac{1}{(2\pi)^{3/2}} \int\limits_{-\infty}^{+\infty} \phi(k) e^{ik \cdot r} d^3k$$

So what is $\phi(k)$? It's the three-dimensional analog of $\phi(k)$ that you find in Chapter 3; that is, it's the amplitude of each component wave function. You can find $\phi(k)$ from the Fourier transform of $\psi_1(x) = Ae^{ik_1 x} + Be^{-ik_1 x}$ (where $x < 0$) like this:

$$\phi(k) = \frac{1}{(2\pi)^{3/2}} \int\limits_{-\infty}^{+\infty} \psi(r) e^{ik \cdot r} d^3k$$

In practice, you choose $\phi(k)$ yourself. Look at an example, using the following form for $\phi(k)$, which is for a Gaussian wave packet (***Note:*** The exponential part is what makes this a Gaussian wave form):

$$\phi(k) = A \exp\left(\frac{-a^2 k^2}{4}\right)$$

where $a$ and A are constants. You can begin by normalizing $\phi(\boldsymbol{k})$ to determine what A is. Here's how that works:

$$1 = \int_{-\infty}^{+\infty} \left|\phi(\boldsymbol{k})\right|^2 d^3k$$

$$1 = \left|A\right|^2 \int_{-\infty}^{+\infty} \exp\left(\frac{-a^2}{2}\boldsymbol{k}^2\right) d^3k$$

Okay. Performing the integral gives you

$$1 = \left|A\right|^2 \left(\frac{2\pi}{a^2}\right)^{3/2}$$

$$A = \left(\frac{a^2}{2\pi}\right)^{3/4}$$

which means that the wave function is

$$\psi(\boldsymbol{r},t) = \frac{1}{\left(2\pi\right)^{3/2}} \left(\frac{a^2}{2\pi}\right)^{3/2} \int_{-\infty}^{+\infty} \exp\left(\frac{-a^2\boldsymbol{k}^2}{4}\right) e^{ik\cdot r} d^3k$$

You can evaluate this equation to give you the following, which is what the time-independent wave function for a Gaussian wave packet looks like in 3D:

$$\psi(\boldsymbol{r},t) = \left(\frac{2}{\pi a^2}\right)^{3/4} \exp\left(\frac{-\boldsymbol{r}^2}{a^2}\right)$$

Okay, that's how things look when $V(\boldsymbol{r}) = 0$. But can't you solve some problems when $V(\boldsymbol{r})$ is not equal to zero? Yep, you sure can. Check out the next section.

# Getting Squared Away with 3D Rectangular Potentials

This section takes a look at a 3D potential that forms a box, as you see in Figure 7-2. You want to get the wave functions and the energy levels here.

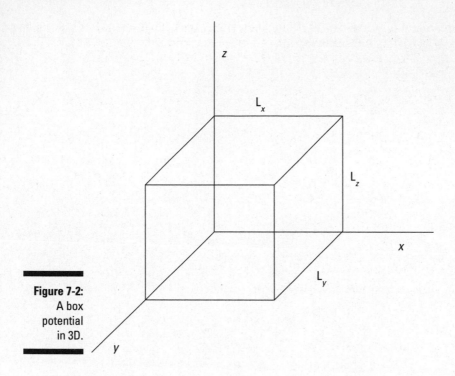

**Figure 7-2:**
A box
potential
in 3D.

Inside the box, say that V(x, y, z) = 0, and outside the box, say that V(x, y, z) = ∞. So you have the following:

$$V(x,y,z) = \begin{vmatrix} 0, \text{ where } 0 < x < L_x, \ 0 < y < L_y, \ 0 < z < L_z \\ \infty \text{ otherwise} \end{vmatrix}$$

Dividing V(x, y, z) into $V_x(x)$, $V_y(y)$, and $V_z(z)$ gives you

↙ $V_x(x) = \begin{vmatrix} 0, \text{ where } 0 < x < L_x \\ \infty \text{ otherwise} \end{vmatrix}$

↙ $V_y(y) = \begin{vmatrix} 0, \text{ where } 0 < y < L_y \\ \infty \text{ otherwise} \end{vmatrix}$

↙ $V_z(z) = \begin{vmatrix} 0, \text{ where } 0 < z < L_z \\ \infty \text{ otherwise} \end{vmatrix}$

Okay, because the potential goes to infinity at the walls of the box, the wave function, $\psi(x, y, z)$, must go to zero at the walls, so that's your constraint. In 3D, the Schrödinger equation looks like this in three dimensions:

$$\frac{-\hbar^2}{2m}\nabla^2\psi(x,y,z)+V(x,y,z)\psi(x,y,z)=E\psi(x,y,z)$$

Writing this out gives you the following:

$$\frac{-\hbar^2}{2m}\left(\frac{\partial^2}{\partial x^2}+\frac{\partial^2}{\partial y^2}+\frac{\partial^2}{\partial z^2}\right)\psi(x,y,z)+V(x,y,z)\psi(x,y,z)=E\psi(x,y,z)$$

Take this dimension by dimension. Because the potential is separable, you can write $\psi(x, y, z)$ as $\psi(x, y, z) = X(x)Y(y)Z(z)$. Inside the box, the potential equals zero, so the Schrödinger equation looks like this for $x$, $y$, and $z$:

✔ $\dfrac{-\hbar^2}{2m}\dfrac{\partial^2}{\partial x^2}X(x)=E_xX(x)$

✔ $\dfrac{-\hbar^2}{2m}\dfrac{\partial^2}{\partial y^2}Y(y)=E_yY(y)$

✔ $\dfrac{-\hbar^2}{2m}\dfrac{\partial^2}{\partial z^2}Z(z)=E_zZ(z)$

The next step is to rewrite these equations in terms of the wave number, $k$. Because $k^2 = \dfrac{2mE}{\hbar^2}$, you can write the Schrödinger equations for $x$, $y$, and $z$ as the following equations:

✔ $\dfrac{\partial^2}{\partial x^2}X(x)=-k_x^{\,2}X(x)$

✔ $\dfrac{\partial^2}{\partial y^2}Y(y)=-k_y^{\,2}Y(y)$

✔ $\dfrac{\partial^2}{\partial z^2}Z(z)=-k_z^{\,2}Z(z)$

Start by taking a look at the equation for $x$. Now you have something to work with — a second order differential equation, $\frac{\partial^2}{\partial x^2}X(x) = -k_x^2 X(x)$ . Here are the two independent solutions to this equation, where A and B are yet to be determined:

- $X_1(x) = A \sin(kx)$
- $X_2(x) = B \cos(kx)$

So the general solution of $\frac{\partial^2}{\partial x^2}X(x) = -k_x^2 X(x)$ is the sum of the last two equations:

$$X(x) = A \sin(kx) + B \cos(kx)$$

Great. Now take a look at determining the energy levels.

## Determining the energy levels

To be able to determine the energy levels of a particle in a box potential, you need an exact value for $X(x)$ — not just one of the terms of the constants A and B. You have to use the boundary conditions to find A and B. What are the boundary conditions? The wave function must disappear at the boundaries of the box, so

- $X(0) = 0$
- $X(L_x) = 0$

So the fact that $\psi(0) = 0$ tells you right away that B must be 0, because $\cos(0) = 1$. And the fact that $X(L_x) = 0$ tells you that $X(L_x) = A \sin(k_x L_x) = 0$. Because the sine is 0 when its argument is a multiple of $\pi$, this means that

$$k_x L_x = n_x \pi \qquad n_x = 1,2,3\dots$$

$$k_x = \frac{n_x \pi}{L_x}$$

And because $k^2 = \frac{2mE}{\hbar^2}$, it means that

$$\frac{2mE_x}{\hbar^2} = \frac{n_x^2 \pi^2}{L_x^2} \qquad n_x = 1,2,3\dots$$

$$E_x = \frac{n_x^2 \hbar^2 \pi^2}{2mL_x^2}$$

That's the energy in the $x$ component of the wave function, corresponding to the quantum numbers 1, 2, 3, and so on. The total energy of a particle of mass $m$ inside the box potential is $E = E_x + E_y + E_z$. Following $E_x = \dfrac{n_x^2 \hbar^2 \pi^2}{2mL_x^2}$, you have this for $E_y$ and $E_z$:

$$E_y = \frac{n_y^2 \hbar^2 \pi^2}{2mL_y^2} \qquad n_y = 1,2,3\ldots$$

$$E_z = \frac{n_z^2 \hbar^2 \pi^2}{2mL_z^2} \qquad n_z = 1,2,3\ldots$$

So the total energy of the particle is $E = E_x + E_y + E_z$, which equals this:

$$E =$$

$$\frac{n_x^2 \hbar^2 \pi^2}{2mL_x^2} \qquad n_x = 1,2,3\ldots$$

$$+\frac{n_y^2 \hbar^2 \pi^2}{2mL_y^2} \qquad n_y = 1,2,3\ldots$$

$$+\frac{n_z^2 \hbar^2 \pi^2}{2mL_z^2} \qquad n_z = 1,2,3\ldots$$

And there you have the total energy of a particle in the box potential.

## Normalizing the wave function

Now how about normalizing the wave function $\psi(x, y, z)$? In the $x$ dimension, you have this for the wave equation:

$$X(x) = A\sin\left(\frac{n_x \pi x}{L_x}\right)$$

So the wave function is a sine wave, going to zero at $x = 0$ and $x = L_z$. You can also insist that the wave function be normalized, like this:

$$1 = \int_0^{L_x} \left|X(x)\right|^2 dx$$

By normalizing the wave function, you can solve for the unknown constant A. Substituting for X(x) in the equation gives you the following:

$$1 = |A|^2 \int_0^{L_x} \sin^2\left(\frac{n_x \pi x}{L_x}\right) dx$$

$$\int_0^{L_x} \sin^2\left(\frac{n_x \pi x}{L_x}\right) dx = \frac{L_x}{2}$$

Therefore, $1 = |A|^2 \int_0^{L_x} \sin^2\left(\frac{n_x \pi x}{L_x}\right) dx$ becomes $1 = |A|^2 \frac{L_x}{2}$, which means you can solve for A:

$$A = \left(\frac{2}{L_n}\right)^{1/2}$$

Great, now you have the constant A, so you can get X(x):

$$X(x) = \left(\frac{2}{L_x}\right)^{1/2} \sin\left(\frac{n_x \pi x}{L_x}\right) \quad n_x = 1,2,3\ldots$$

Now get ψ(x, y, z). You can divide the wave function into three parts:

ψ(x, y, z) = X(x)Y(y)Z(z)

By analogy with X(x), you can find Y(y) and Z(z):

$$Y(y) = \left(\frac{2}{L_y}\right)^{1/2} \sin\left(\frac{n_y \pi y}{L_y}\right) \quad n_y = 1,2,3\ldots$$

$$Z(z) = \left(\frac{2}{L_z}\right)^{1/2} \sin\left(\frac{n_z \pi z}{L_z}\right) \quad n_z = 1,2,3\ldots$$

So ψ(x, y, z) equals the following:

$$\psi(x,y,z) = \left(\frac{8}{L_x L_y L_z}\right)^{1/2} \sin\left(\frac{n_x \pi x}{L_x}\right) \sin\left(\frac{n_y \pi x}{L_y}\right) \sin\left(\frac{n_z \pi x}{L_z}\right)$$

$$n_x = 1,2,3\ldots$$
$$n_y = 1,2,3\ldots$$
$$n_z = 1,2,3\ldots$$

That's a pretty long wave function. In fact, when you're dealing with a box potential, the energy looks like this:

$$E =$$

$$\frac{n_x^2 \hbar^2 \pi^2}{2mL_x^2} \qquad n_x = 1,2,3\ldots$$

$$+\frac{n_y^2 \hbar^2 \pi^2}{2mL_y^2} \qquad n_y = 1,2,3\ldots$$

$$+\frac{n_z^2 \hbar^2 \pi^2}{2mL_z^2} \qquad n_z = 1,2,3\ldots$$

## Using a cubic potential

When working with a box potential, you can make things simpler by assuming that the box is actually a cube. In other words, $L = L_x = L_y = L_z$. When the box is a cube, the equation for the energy becomes

$$E = \frac{\hbar^2 \pi^2}{2mL^2}\left(n_x^2 + n_y^2 + n_z^2\right)$$

$$n_x = 1,2,3\ldots$$

$$n_y = 1,2,3\ldots$$

$$n_z = 1,2,3\ldots$$

So, for example, the energy of the ground state, where $n_x = n_y = n_z = 1$, is given by the following, where $E_{111}$ is the ground state:

$$E_{111} = \frac{3\hbar^2 \pi^2}{2mL^2}$$

Note that there's some degeneracy in the energies; for example, note that

✔ $E_{211}$ ($n_x = 2$, $n_y = 1$, $n_z = 1$) is $E_{211} = \frac{6\hbar^2 \pi^2}{2mL^2}$

✔ $E_{121}$ ($n_x = 1$, $n_y = 2$, $n_z = 1$) is $E_{121} = \frac{6\hbar^2 \pi^2}{2mL^2}$

✔ $E_{112}$ ($n_x = 1$, $n_y = 1$, $n_z = 2$) is $E_{112} = \frac{6\hbar^2 \pi^2}{2mL^2}$

So $E_{211} = E_{121} = E_{112}$, which means that the first excited state is threefold degenerate, matching the threefold equivalence in dimensions.

In general, when you have symmetry built into the physical layout (as you do when $L = L_x = L_y = L_z$), you have degeneracy.

The wave function for a cubic potential is also easier to manage than the wave function for a general box potential (where the sides aren't of the same length). Here's the wave function for a cubic potential:

$$\psi(x,y,z) = \left(\frac{8}{L^3}\right)^{\frac{1}{2}} \sin\left(\frac{n_x \pi x}{L}\right) \sin\left(\frac{n_y \pi y}{L}\right) \sin\left(\frac{n_z \pi z}{L}\right)$$

$$n_x = 1,2,3\ldots$$
$$n_y = 1,2,3\ldots$$
$$n_z = 1,2,3\ldots$$

So, for example, here's the wave function for the ground state ($n_x = 1$, $n_y = 1$, $n_z = 1$), $\psi_{111}(x, y, z)$:

$$\psi_{111}(x,y,z) = \left(\frac{8}{L^3}\right)^{\frac{1}{2}} \sin\left(\frac{\pi x}{L}\right) \sin\left(\frac{\pi y}{L}\right) \sin\left(\frac{\pi z}{L}\right)$$

And here's $\psi_{211}(x, y, z)$:

$$\psi_{111}(x,y,z) = \left(\frac{8}{L^3}\right)^{\frac{1}{2}} \sin\left(\frac{\pi x}{L}\right) \sin\left(\frac{\pi y}{L}\right) \sin\left(\frac{\pi z}{L}\right)$$

And $\psi_{121}(x, y, z)$:

$$\psi_{121}(x,y,z) = \left(\frac{8}{L^3}\right)^{\frac{1}{2}} \sin\left(\frac{\pi x}{L}\right) \sin\left(\frac{\pi y}{L}\right) \sin\left(\frac{\pi z}{L}\right)$$

# Springing into 3D Harmonic Oscillators

In one dimension, the general particle harmonic oscillator (which I first describe in Chapter 4) looks like Figure 7-3, where the particle is under the influence of a restoring force — here illustrated as a spring.

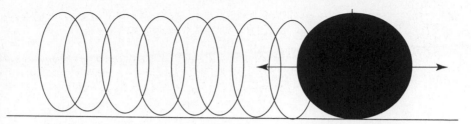

**Figure 7-3:**
A harmonic
oscillator.

The restoring force has the form $F_x = -k_x x$ in one dimension, where $k_x$ is the constant of proportionality between the force on the particle and the location of the particle. The potential energy of the particle as a function of location $x$ is $V(x) = \frac{1}{2}k_x x^2$. This is also sometimes written as

$$V(x) = \frac{1}{2}m\omega_x^2 x^2$$

where $\omega_x^2 = k_x \big/ m$.

In this section, you take a look at the harmonic oscillator in three dimensions. In three dimensions, the potential looks like this:

$$V(x,y,z) = \frac{1}{2}m\omega_x^2 x^2 + \frac{1}{2}m\omega_y^2 y^2 + \frac{1}{2}m\omega_z^2 z^2$$
$$\omega_x^2 = k_x \big/ m$$
$$\omega_y^2 = k_y \big/ m$$
$$\omega_z^2 = k_z \big/ m$$

Now that you have a form for the potential, you can start talking in terms of Schrödinger's equation:

$$\frac{-\hbar^2}{2m}\left(\frac{\partial^2}{\partial x^2} + \frac{\partial^2}{\partial y^2} + \frac{\partial^2}{\partial z^2}\right)\psi(x,y,z) + V(x,y,z)\psi(x,y,z) = E\psi(x,y,z)$$

Substituting in for the three-dimension potential, V(x, y, z), gives you this equation:

$$\frac{-\hbar^2}{2m}\left(\frac{\partial^2}{\partial x^2}+\frac{\partial^2}{\partial y^2}+\frac{\partial^2}{\partial z^2}\right)\psi\left(x,y,z\right)$$

$$+\left(\frac{1}{2}m\omega_x^2x^2+\frac{1}{2}m\omega_y^2y^2+\frac{1}{2}m\omega_z^2z^2\right)\psi\left(x,y,z\right)=E\psi\left(x,y,z\right)$$

Take this dimension by dimension. Because you can separate the potential into three dimensions, you can write ψ(x, y, z) as ψ(x, y, z) = X(x)Y(y)Z(z). Therefore, the Schrödinger equation looks like this for *x*:

$$\frac{-\hbar^2}{2m}\frac{\partial^2}{\partial x^2}X\left(x\right)+\frac{1}{2}m\omega_x^2x^2X\left(x\right)=E_xX\left(x\right)$$

You solve that equation in Chapter 4, where you get this next solution:

$$X\left(x\right)=\frac{1}{\pi^{1/4}\left(2^{n_x}n_x!x_{x0}\right)^{1/2}}H_n\left(\frac{x}{x_0}\right)\exp\left(\frac{-x^2}{2x_{x0}^2}\right)$$

where $X_0=\left(\frac{\hbar}{m\omega_x}\right)^{\frac{1}{2}}$ and $n_x$ = 0, 1, 2, and so on. The $H_n$ term indicates a hermite polynomial, which looks like this:

- ✔ $H_0(x) = 1$
- ✔ $H_1(x) = 2x$
- ✔ $H_2(x) = 4x^2 - 2$
- ✔ $H_3(x) = 8x^3 - 12x$
- ✔ $H_4(x) = 16x^4 - 48x^2 + 12$
- ✔ $H_5(x) = 32x^5 - 160x^3 + 120x$

Therefore, you can write the wave function like this:

$$\psi\left(x,y,z\right)=\frac{\frac{1}{\pi^{3/4}}\dfrac{H_{nx}\left(\frac{x}{x_0}\right)\exp\left(\frac{-x^2}{2x_0^2}\right)}{\left(2^{n_x}n_x!x_{x0}\right)^{1/2}}}{\dfrac{H_{ny}\left(\frac{y}{y_0}\right)\exp\left(\frac{-y^2}{2y_0^2}\right)}{\left(2^{n_y}n_y!y_0\right)^{1/2}}\dfrac{H_{nz}\left(\frac{z}{z_0}\right)\exp\left(\frac{-z^2}{2z_0^2}\right)}{\left(2^{n_z}n_z!z_0\right)^{1/2}}}$$

That's a relatively easy form for a wave function, and it's all made possible by the fact that you can separate the potential into three dimensions.

What about the energy of the harmonic oscillator? The energy of a one-dimensional harmonic oscillator is $E = \left( n + \dfrac{1}{2} \right) \hbar \omega$. And by analogy, the energy of a three-dimensional harmonic oscillator is given by

$$E = \left( n_x + \frac{1}{2} \right) \hbar \omega_x + \left( n_y + \frac{1}{2} \right) \hbar \omega_y + \left( n_z + \frac{1}{2} \right) \hbar \omega_z$$

Note that if you have an isotropic harmonic oscillator, where $\omega_x = \omega_y = \omega_z = \omega$, the energy looks like this:

$$E = \left( n_x + n_y + n_z + \frac{3}{2} \right) \hbar \omega$$

As for the cubic potential, the energy of a 3D isotropic harmonic oscillator is degenerate. For example, $E_{112} = E_{121} = E_{211}$. In fact, it's possible to have more than threefold degeneracy for a 3D isotropic harmonic oscillator — for example, $E_{200} = E_{020} = E_{002} = E_{110} = E_{101} = E_{011}$.

In general, the degeneracy of a 3D isotropic harmonic oscillator is

$$\text{Degeneracy} = \frac{1}{2} (n+1)(n+2)$$

where $n = n_x + n_y + n_z$.

# Chapter 8

# Solving Problems in Three Dimensions: Spherical Coordinates

*I*n your other life as a sea captain-slash-pilot, you're probably pretty familiar with latitude and longitude — coordinates that basically name a couple of angles as measured from the center of the Earth. Put together the angle east or west, the angle north or south, and the all-important distance from the center of the Earth, and you have a vector that gives a good description of location in three dimensions. That vector is part of a *spherical coordinate system.*

Navigators talk more about the pair of angles than the distance ("Earth's surface" is generally specific enough for them), but quantum physicists find both angles and radius length important. Some 3D quantum physics problems even allow you to break down a wave function into two parts: an angular part and a radial part.

In this chapter, I discuss three-dimensional problems that are best handled using spherical coordinates. (For 3D problems that work better in rectangular coordinate systems, see Chapter 7.)

# A New Angle: Choosing Spherical Coordinates Instead of Rectangular

Say you have a 3D box potential, and suppose that the potential well that the particle is trapped in looks like this, which is suited to working with rectangular coordinates:

$$V(x,y,z) = \begin{vmatrix} 0, \text{ where } 0 < x < L_x, \ 0 < y < L_y, \ 0 < z < L_z \\ \infty \text{ otherwise} \end{vmatrix}$$

Because you can easily break this potential down in the $x$, $y$, and $z$ directions, you can break the wave function down that way, too, as you see here:

$$\psi(x, y, z) = X(x)Y(y)Z(z)$$

Solving for the wave function gives you the following normalized result in rectangular coordinates:

$$\psi(x,y,z) = \left( \frac{8}{L_x L_y L_z} \right)^{\frac{1}{2}} \sin \frac{(n_x \pi x)}{L_x} \sin \frac{(n_y \pi y)}{L_y} \sin \frac{(n_z \pi z)}{L_z}$$

$$n_x = 1,2,3\ldots$$
$$n_y = 1,2,3\ldots$$
$$n_z = 1,2,3\ldots$$

The energy levels also break down into separate contributions from all three rectangular axes:

$$E = E_x + E_y + E_z$$

And solving for E gives you this equation (from Chapter 7):

$$E =$$
$$\frac{n_x^2 \hbar^2 \pi^2}{2mL_x^2} \qquad n_x = 1,2,3\ldots$$
$$+\frac{n_y^2 \hbar^2 \pi^2}{2mL_y^2} \qquad n_y = 1,2,3\ldots$$
$$+\frac{n_z^2 \hbar^2 \pi^2}{2mL_z^2} \qquad n_z = 1,2,3\ldots$$

But what if the potential well a particle is trapped in has spherical symmetry, not rectangular? For example, what if the potential well were to look like this, where $r$ is the radius of the particle's location with respect to the origin and where $a$ is a constant?

$$V(r) = \begin{cases} 0, \text{ where } 0 < r < a \\ \infty \text{ otherwise} \end{cases}$$

Clearly, trying to stuff this kind of problem into a rectangular-coordinates kind of solution is only asking for trouble, because although you can do it, it involves lots of sines and cosines and results in a pretty complex solution. A much better tactic is to solve this kind of a problem in the natural coordinate system in which the potential is expressed: spherical coordinates.

Figure 8-1 shows the spherical coordinate system along with the corresponding rectangular coordinates, $x$, $y$, and $z$. In the spherical coordinate system, you locate points with a radius vector named **r,** which has three components:

  ✔ An $r$ component (the length of the radius vector)

  ✔ $\theta$ (the angle from $z$ axis to the the **r** vector)

  ✔ $\phi$ (the angle from the $x$ axis to the the **r** vector)

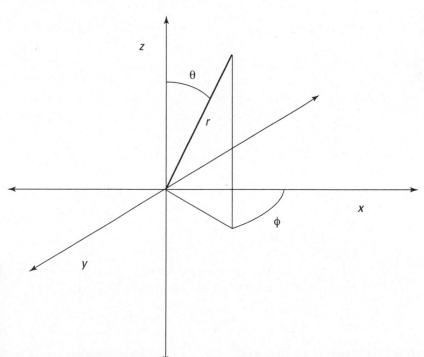

**Figure 8-1:**
The
spherical
coordinate
system.

# Taking a Good Look at Central Potentials in 3D

This chapter focuses on problems that involve *central potentials* — that is, spherically symmetrical potentials, of the kind where $V(\mathbf{r}) = V(r)$. In other words, the potential is independent of the vector nature of the radius vector; the potential depends on only the magnitude of vector $\mathbf{r}$ (which is $r$), not on the angle of $\mathbf{r}$.

When you work on problems that have a central potential, you're able to separate the wave function into a radial part (which depends on the form of the potential) and an angular part, which is a spherical harmonic. Read on.

## Breaking down the Schrödinger equation

The Schrödinger equation looks like this in three dimensions, where $\Delta$ is the Laplacian operator (see Chapter 2 for more on operators):

$$\frac{-\hbar^2}{2m}\Delta\psi(\mathbf{r})+V(\mathbf{r})\psi(\mathbf{r})=E\psi(\mathbf{r})$$

And the Laplacian operator looks like this in rectangular coordinates:

$$\Delta=\frac{\partial^2}{\partial x^2}+\frac{\partial^2}{\partial y^2}+\frac{\partial^2}{\partial z^2}$$

In spherical coordinates, it's a little messy, but you can simplify later. Check out the spherical Laplacian operator:

$$\Delta=\frac{1}{r}\frac{\partial^2}{\partial r^2}r-\frac{1}{\hbar^2 r^2}\mathbf{L}^2$$

Here, $\mathbf{L}^2$ is the square of the orbital angular momentum:

$$\mathbf{L}^2=-\hbar^2\left(\frac{1}{\sin\theta}\frac{\partial}{\partial\theta}\left(\sin\theta\frac{\partial}{\partial\theta}\right)+\frac{1}{\sin^2\theta}\frac{\partial^2}{\partial\phi^2}\right)$$

So in spherical coordinates, the Schrödinger equation for a central potential looks like this when you substitute in the terms:

$$\frac{-\hbar^2}{2m}\frac{1}{r}\frac{\partial^2}{\partial r^2}\psi(\mathbf{r})+\frac{1}{2mr^2}\mathbf{L}^2\psi(\mathbf{r})+V(r)\psi(\mathbf{r})=E\psi(\mathbf{r})$$

REMEMBER

Take a look at the preceding equation. The first term actually corresponds to the *radial kinetic energy* — that is, the kinetic energy of the particle moving in the radial direction. The second term corresponds to the *rotational kinetic energy*. And the third term corresponds to the *potential energy*.

So what can you say about the solutions to this version of the Schrödinger equation? You can note that the first term depends only on $r$, as does the third, and that the second term depends only on angles. So you can break the wave function, $\psi(r) = \psi(r, \theta, \phi)$, into two parts:

▸ A radial part

▸ A part that depends on the angles

This is a special property of problems with central potentials.

## The angular part of $\psi(r, \theta, \phi)$

When you have a central potential, what can you say about the angular part of $\psi(r, \theta, \phi)$? The angular part must be an eigenfunction of $\mathbf{L}^2$, and as I show in Chapter 5, the eigenfunctions of $\mathbf{L}^2$ are the spherical harmonics, $Y_{lm}(\theta, \phi)$ (where $l$ is the total angular momentum quantum number and $m$ is the $z$ component of the angular momentum's quantum number). The spherical harmonics equal

$$Y_{lm}(\theta,\phi) = (-1)^m \left[ \frac{(2l+1)(l-m)!}{4\pi(l+m)!} \right]^{\frac{1}{2}} P_{lm}(\cos\theta) e^{im\phi} \quad \text{where } m \geq 0$$

Here are the first several normalized spherical harmonics:

▸ $Y_{00}(\theta,\phi) = \dfrac{1}{(4\pi)^{\frac{1}{2}}}$

▸ $Y_{10}(\theta, \phi) = \left( \dfrac{3}{4\pi} \right)^{\frac{1}{2}} \cos\theta$

▸ $Y_{1\pm1}(\theta,\phi) = \pm\left( \dfrac{3}{8\pi} \right)^{\frac{1}{2}} e^{\pm i\phi} \sin\theta$

▸ $Y_{20}(\theta, \phi) = \left( \dfrac{5}{16\pi} \right)^{\frac{1}{2}} (3\cos^2\theta - 1)$

▸ $Y_{2\pm1}(\theta, \phi) = \pm\left( \dfrac{15}{8\pi} \right)^{\frac{1}{2}} e^{\pm i\phi} \sin\theta$

▸ $Y_{2\pm2}(\theta, \phi) = \left( \dfrac{15}{32\pi} \right)^{\frac{1}{2}} e^{\pm 2i\phi} \sin^2\theta$

That's what the angular part of the wave function is going to be: a spherical harmonic.

## The radial part of $\psi(r, \theta, \phi)$

You can give the radial part of the wave function the name $R_{nl}(r)$, where $n$ is a quantum number corresponding to the quantum state of the radial part of the wave function and $l$ is the total angular momentum quantum number. The radial part is symmetric with respect to angles, so it can't depend on $m$, the quantum number of the $z$ component of the angular momentum. In other words, the wave function for particles in central potentials looks like the following equation in spherical coordinates:

$$\psi(r, \theta, \phi) = R_{nl}(r)Y_{lm}(\theta, \phi)$$

The next step is to solve for $R_{nl}(r)$ in general. Substituting $\psi(r, \theta, \phi)$ from the preceding equation into the Schrödinger equation,
$\dfrac{-\hbar^2}{2m}\dfrac{1}{r}\dfrac{\partial^2}{\partial r^2}\psi(r) + \dfrac{1}{2mr^2}\mathbf{L}^2\psi(r) + V(r)\psi(r) = E\psi(r)$, gives you

$$-\hbar^2\frac{r}{R_{nl}(r)}\frac{\partial^2}{\partial r^2}\Big[r\,R_{nl}(r)\Big] + 2mr^2\Big[V(r) - E\Big] + \frac{\mathbf{L}^2 Y_{lm}(\theta,\phi)}{Y_{lm}(\theta,\phi)} = 0$$

Okay, what can you make of this? First, note (from Chapter 5) that the spherical harmonics are eigenfunctions of $\mathbf{L}^2$ (that's the whole reason for using them), with eigenvalue $l(l + 1)\hbar^2$:

$$\mathbf{L}^2 Y_{lm}(\theta,\phi) = l(l+1)\hbar^2 Y_{lm}(\theta,\phi)$$

So the last term in this equation is simply $l(l + 1)\hbar^2$. That means that

$$-\hbar^2\frac{r}{R_{nl}(r)}\frac{\partial^2}{\partial r^2}\Big[r\,R_{nl}(r)\Big] + 2mr^2\Big[V(r) - E\Big] + \frac{\mathbf{L}^2 Y_{lm}(\theta,\phi)}{Y_{lm}(\theta,\phi)} = 0 \text{ takes the form}$$

$$-\hbar^2\frac{r}{R_{nl}(r)}\frac{\partial^2}{\partial r^2}\Big[r\,R_{nl}(r)\Big] + 2mr^2\Big[V(r) - E\Big] + l(l+1)\hbar^2 = 0 \text{, which equals}$$

$$\frac{-\hbar^2}{2m}\frac{d^2}{dr^2}\Big[r\,R_{nl}(r)\Big] + \left[V(r) + \frac{l(l+1)\hbar^2}{2mr^2}\right]\Big[r\,R_{nl}(r)\Big] = E\Big[r\,R_{nl}(r)\Big]$$

The preceding equation is the one you use to determine the radial part of the wave function, $R_{nl}(r)$. It's called the *radial equation* for a central potential.

When you solve the radial equation for $R_{nl}(r)$, you can then find $\psi(r, \theta, \phi)$ because you already know $Y_{lm}(\theta,\phi)$:

$\psi(r, \theta, \phi) = R_{nl}(r)Y_{lm}(\theta,\phi)$

Thus, this chapter simply breaks down to finding the solution to the radial equation.

**Note:** Incidentally, the radial equation is really a differential equation in one dimension: the $r$ dimension. By selecting only problems that contain central potentials, you reduce the general problem of finding the wave function of particles trapped in a three-dimensional spherical potential to a one-dimensional differential equation.

# Handling Free Particles in 3D with Spherical Coordinates

In this section and the next, you take a look at some example central potentials to see how to solve the radial equation (see the preceding section for more on the radial part). Here, you work with a free particle, in which no potential at all constrains the particle.

The wave function in spherical coordinates takes this form:

$$\psi(r,\theta,\phi) = R_{nl}(r)Y_{lm}(\theta,\phi)$$

And you know all about $Y_{lm}(\theta, \phi)$, because it gives you the spherical harmonics. The problem is now to solve for the radial part, $R_{nl}(r)$. Here's the radial equation:

$$\frac{-\hbar^2}{2m}\frac{\partial^2}{\partial r^2}\big[r\,R_{nl}(r)\big] + \left[V(r) + \frac{l(l+1)\hbar^2}{2mr^2}\right]\big[r\,R_{nl}(r)\big] = E\big[r\,R_{nl}(r)\big]$$

For a free particle, $V(r) = 0$, so the radial equation becomes

$$\frac{-\hbar^2}{2m}\frac{\partial^2}{\partial r^2}\big[r\,R_{nl}(r)\big] + \frac{l(l+1)\hbar^2}{2mr^2}\big[r\,R_{nl}(r)\big] = E\big[r\,R_{nl}(r)\big]$$

The way you usually handle this equation is to substitute $\rho$ for $kr$, where $k = (2mE_n)^{1/2}/\hbar$, so that $R_{nl}(r)$ becomes $R_l(kr) = R_l(\rho)$. This substitution means that

$$\frac{-\hbar^2}{2m}\frac{\partial^2}{\partial r^2}\left[r\, R_{nl}(r)\right] + \frac{l(l+1)\hbar^2}{2mr^2}\left[r\, R_{nl}(r)\right] = E\left[r\, R_{nl}(r)\right]$$ becomes the following:

$$\frac{d^2 R_l(\rho)}{d\rho^2} + \frac{2}{\rho}\frac{dR_l(\rho)}{d\rho} + \left[1 - \frac{l(l+1)}{\rho^2}\right]R_l(\rho) = 0$$

In this section, you see how the spherical Bessel and Neumann functions come to the rescue when you're dealing with free particles.

## The spherical Bessel and Neumann functions

The radial part of the equation, $\dfrac{d^2 R_l(\rho)}{d\rho^2} + \dfrac{2}{\rho}\dfrac{dR_l(\rho)}{d\rho} + \left[1 - \dfrac{l(l+1)}{\rho^2}\right]R_l(\rho) = 0$,

looks tough, but the solutions turn out to be well-known — this equation is called the *spherical Bessel equation,* and the solution is a combination of the spherical Bessel functions [$j_l(\rho)$] and the spherical Neumann functions [$n_l(\rho)$]:

$$R_l(\rho) = A_l j_l(\rho) + B_l n_l(\rho)$$

So what are the spherical Bessel functions and the spherical Neumann functions? The spherical Bessel functions are given by

$$j_l(\rho) = (-\rho)^l\left(\frac{1}{\rho}\frac{d}{d\rho}\right)^l\frac{\sin\rho}{\rho}$$

Here's what the first few iterations of $j_l(\rho)$ look like:

- $j_0(\rho) = \dfrac{\sin\rho}{\rho}$

- $j_1(\rho) = \dfrac{\sin\rho}{\rho^2} - \dfrac{\cos\rho}{\rho}$

- $j_2(\rho) = \dfrac{3\sin\rho}{\rho^3} - \dfrac{3\cos\rho}{\rho^2} - \dfrac{\sin\rho}{\rho}$

How about the spherical Neumann functions? The spherical Neumann functions are given by

$$n_l(\rho) = -(-\rho)^l\left(\frac{1}{\rho}\frac{d}{d\rho}\right)^l\frac{\cos\rho}{\rho}$$

Here are the first few iterations of $n_l(\rho)$:

- $n_0(\rho) = -\dfrac{\cos\rho}{\rho}$

- $n_1(\rho) = -\dfrac{\cos\rho}{\rho^2} - \dfrac{\sin\rho}{\rho}$

- $n_2(\rho) = -\dfrac{3\cos\rho}{\rho^3} - \dfrac{3\sin\rho}{\rho^2} + \dfrac{\cos\rho}{\rho}$

## The limits for small and large ρ

According to the spherical Bessel equation, the radial part of the wave function for a free particle looks like this:

$$R_l(\rho) = A_l j_l(\rho) + B_l n_l(\rho)$$

Take a look at the spherical Bessel functions and Neumann functions for small and large ρ:

- **Small ρ:** The Bessel functions reduce to $j_l(\rho) \approx \dfrac{2^l l! \rho^l}{(2^l + 1)!}$

  The Neumann functions reduce to $n_l(\rho) \approx \dfrac{-(2l-1)! \rho^{-l-1}}{2^l l!}$

- **Large ρ:** The Bessel functions reduce to $j_l(\rho) \approx \dfrac{1}{\rho}\sin\left(\rho - \dfrac{l\pi}{2}\right)$

  The Neumann functions reduce to $n_l(\rho) \approx -\dfrac{1}{\rho}\cos\left(\rho - \dfrac{l\pi}{2}\right)$.

Note that the Neumann functions diverge for small ρ. Therefore, any wave function that includes the Neumann functions also diverges, which is unphysical. So the Neumann functions aren't acceptable functions in the wave function.

That means the wave function $\psi(r, \theta, \phi)$, which equals $R_{nl}(r)\, Y_{lm}(\theta, \phi)$, equals the following:

$$\psi(r, \theta, \phi) = j_l(kr)\, Y_{lm}(\theta, \phi)$$

where $k = (2mE_n)^{1/2}/\hbar$. Note that because $k$ can take any value, the energy levels are continuous.

# Handling the Spherical Square Well Potential

Take a look at a spherical square well potential of the kind you can see in Figure 8-2 (I introduce square wells in Chapter 3). This potential traps particles inside it. Mathematically, you can express the square well potential like this:

$$V(r) = \begin{vmatrix} -V_0, \text{ where } 0 < r < a \\ 0, \text{ where } r > a \end{vmatrix}$$

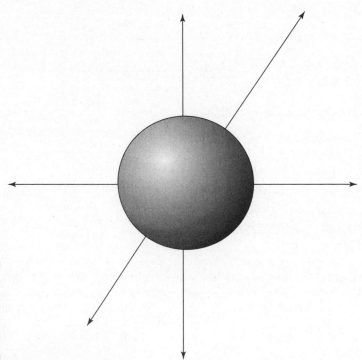

**Figure 8-2:**
The
spherical
square well
potential.

Note that this potential is spherically symmetric and varies only in $r$, not in $\theta$ or $\phi$. You're dealing with a central potential, so you can break the wave function into an angular part and a radial part (see the earlier section "Taking a Good Look at Central Potentials in 3D").

This section has you take a look at the radial equation, handling the two cases of $0 < r < a$ and $r > a$ separately.

# Inside the square well: $0 < r < a$

For a spherical square well potential, here's what the radial equation looks like for the region $0 < r < a$:

$$\frac{-\hbar^2}{2m}\frac{\partial^2}{\partial r^2}\left[r\,R_{nl}(r)\right]+\left[V(r)+\frac{l(l+1)\hbar^2}{2mr^2}\right]\left[r\,R_{nl}(r)\right]=E\left[r\,R_{nl}(r)\right]$$

In this region, $V(r) = -V_0$, so you have

$$\frac{-\hbar^2}{2m}\frac{\partial^2}{\partial r^2}\left[r\,R_{nl}(r)\right]+\left[-V_0+\frac{l(l+1)\hbar^2}{2mr^2}\right]\left[r\,R_{nl}(r)\right]=E\left[r\,R_{nl}(r)\right]$$

Taking the $V_0$ term over to the right gives you the following:

$$\frac{-\hbar^2}{2m}\frac{\partial^2}{\partial r^2}\left[r\,R_{nl}(r)\right]+\frac{l(l+1)\hbar^2}{2mr^2}\left[r\,R_{nl}(r)\right]=(E+V_0)r\,R_{nl}(r)$$

And here's what dividing by $r$ gives you:

$$\frac{-\hbar^2}{2m}\frac{1}{r}\frac{\partial^2}{\partial r^2}\left[r\,R_{nl}(r)\right]+\frac{l(l+1)\hbar^2}{2mr^2}R_{nl}(r)=(E+V_0)R_{nl}(r)$$

Then, multiplying by $-2m/\hbar^2$, you get

$$\frac{1}{r}\frac{\partial^2}{\partial r^2}\left[r\,R_{nl}(r)\right]-\frac{l(l+1)}{r^2}R_{nl}(r)=-\frac{2m}{\hbar^2}(E+V_0)R_{nl}(r)$$

Now make the change of variable $\rho = kr$, where $k = (2m(E+V_0))^{1/2}/\hbar$, so that $R_{nl}(r)$ becomes $R_l(kr) = R_l(\rho)$. Using this substitution means that $V(r)=\begin{vmatrix}-V_0,\text{ where }0<r<a\\0,\text{ where }r>a\end{vmatrix}$ takes the following form:

$$\frac{d^2R_l(\rho)}{d\rho^2}+\frac{2}{\rho}\frac{dR_l(\rho)}{d\rho}+\left[1-\frac{l(l+1)}{\rho^2}\right]R_l(\rho)=0$$

This is the spherical Bessel equation (just as you see for the free particle in "Handling Free Particles in 3D with Spherical Coordinates"). This time, $k = [2m(E+V_0)]^{1/2}/\hbar$, not $(2mE)^{1/2}/\hbar$. That makes sense, because now the particle is trapped in the square well, so its total energy is $E + V_0$, not just E.

The solution to the preceding equation is a combination of the spherical Bessel functions [$j_l(\rho)$] and the spherical Neumann functions [$n_l(\rho)$]:

$$R_l(\rho) = A_l j_l(\rho) + B_l n_l(\rho)$$

You can apply the same constraint here that you apply for a free particle: The wave function must be finite everywhere. For small $\rho$, the Bessel functions look like this:

$$j_l(\rho) \approx \frac{2^l l! \rho^l}{(2^l + 1)!}$$

And for small $\rho$, the Neumann functions reduce to

$$n_l(\rho) \approx \frac{-(2l-1)! \rho^{-l-1}}{2^l l!}$$

So the Neumann functions diverge for small $\rho$, which makes them unacceptable for wave functions here. That means that the radial part of the wave function is just made up of spherical Bessel functions, where $A_l$ is a constant:

$$R_l(\rho) = A_l j_l(\rho)$$

The whole wave function inside the square well, $\psi_{inside}(r, \theta, \phi)$, is a product of radial and angular parts, and it looks like this:

$$\psi_{inside}(r, \theta, \phi) = A_l j_l(\rho_{inside}) Y_{lm}(\theta, \phi)$$

where $\rho_{inside} = r(2m(E + V_0))^{1/2}/\hbar$ and $Y_{lm}(\theta, \phi)$ are the spherical harmonics.

## Outside the square well: r > a

Outside the square well, in the region $r > a$, the particle is just like a free particle, so here's what the radial equation looks like:

$$\frac{-\hbar^2}{2m} \frac{\partial^2}{\partial r^2} \left[ r\, R_{nl}(r) \right] + \left[ V(r) + \frac{l(l+1)\hbar^2}{2mr^2} \right] \left[ r\, R_{nl}(r) \right] = E\, r\, R_{nl}(r)$$

You solve this equation earlier in "Handling Free Particles in 3D with Spherical Coordinates": Because $\rho = kr$, where $k = (2mE)^{1/2}/\hbar$, you substitute $\rho$ for $kr$ so that $R_{nl}(r)$ becomes $R_l(kr) = R_l(\rho)$. Using this substitution means that the radial equation takes the following form:

$$\frac{d^2R_l(\rho)}{d\rho^2} + \frac{2}{\rho}\frac{dR_l(\rho)}{d\rho} + \left[1 - \frac{l(l+1)}{\rho^2}\right]R_l(\rho) = 0$$

The solution is a combination of spherical Bessel functions and spherical Neumann functions, where $B_l$ is a constant:

$$R_l(\rho) = B_l[j_l(\rho) + n_l(\rho)]$$

So the radial solution outside the square well looks like this, where $\rho_{outside} = r(2mE)^{1/2}/\hbar$:

$$\psi_{outside}(r,\theta,\phi) = B_l\left[j_l(\rho_{outside}) + n_l(\rho_{outside})\right]Y_{lm}(\theta,\phi)$$

From the preceding section, you know that the wave function inside the square well is

$$\psi_{inside}(r,\theta,\phi) = A_l j_l(\rho_{inside})Y_{lm}(\theta,\phi)$$

So how do you find the constants $A_l$ and $B_l$? You find those constants through continuity constraints: At the inside/outside boundary, where $r = a$, the wave function and its first derivative must be continuous. So to determine $A_l$ and $B_l$, you have to solve these two equations:

$$\boldsymbol{\checkmark}\ \psi_{inside}(a,\theta,\phi) = \psi_{outside}(a,\theta,\phi)$$

$$\boldsymbol{\checkmark}\ \frac{d}{dr}\psi_{inside}(r,\theta,\phi)\bigg|_{r=a} = \frac{d}{dr}\psi_{outside}(r,\theta,\phi)\bigg|_{r=a}$$

# Getting the Goods on Isotropic Harmonic Oscillators

This section takes a look at spherically symmetric harmonic oscillators in three dimensions. In one dimension, you write the harmonic oscillator potential like this:

$$V(x) = \frac{1}{2}m\omega^2 x^2$$

where $\omega^2 = \frac{k}{m}$ (here, $k$ is the spring constant; that is, the restoring force of the harmonic oscillator is $F = -kx$). You can turn these two equations into three-dimensional versions of the harmonic potential by replacing $x$ with $r$:

$$V(r) = \frac{1}{2}m\omega^2 r^2$$

where $\omega^2 = \frac{k}{m}$. Because this potential is spherically symmetric, the wave function is going to be of the following form:

$$\psi(r,\theta,\phi) = R_{nl}(r)Y_{lm}(\theta,\phi)$$

where you have yet to solve for the radial function $R_{nl}(r)$ and where $Y_{lm}(\theta, \phi)$ describes the spherical harmonics.

The Schrödinger equation looks like this in three dimensions:

$$\frac{-\hbar^2}{2m}\frac{\partial^2}{\partial r^2}\left[r\,R_{nl}(r)\right] + \left[V(r) + \frac{l(l+1)\hbar^2}{2mr^2}\right]\left[r\,R_{nl}(r)\right] = E\left[r\,R_{nl}(r)\right]$$

Substituting for V(r) from $V(r) = \frac{1}{2}m\omega^2 r^2$ gives you the following:

$$\frac{-\hbar^2}{2m}\frac{\partial^2}{\partial r^2}\left[r\,R_{nl}(r)\right] + \left[\frac{1}{2}m\omega^2 r^2 + \frac{l(l+1)\hbar^2}{2mr^2}\right]\left[r\,R_{nl}(r)\right] = E\left[r\,R_{nl}(r)\right]$$

Well, the solution to this equation is pretty difficult to obtain, and you're not going to gain anything by going through the math (pages and pages of it), so here's the solution:

$$R_{nl}(r) = C_{nl}r^l \exp\left(-m\omega\frac{r^2}{2\hbar}\right)L_{\frac{(n-1)}{2}}^{\left(l+\frac{1}{2}\right)}\left(m\omega\frac{r^2}{\hbar}\right)$$

where $\exp(x) = e^x$ and

$$C_{nl} = \frac{\left[\dfrac{2^{n+l+2}\left(\dfrac{m\omega}{\hbar}\right)^{l+\frac{3}{2}}}{\pi^{\frac{1}{2}}}\right]^{\frac{1}{2}}\left[\dfrac{n-l}{2}\right]!\left[\dfrac{n+l}{2}\right]!}{\left[(n+l+1)!\right]^{\frac{1}{2}}}$$

And the $L_a^b(r)$ functions are the generalized Laguerre polynomials:

$$L_a^b(r) = \frac{r^{-b}e^r}{a!}\frac{d^a}{dr^a}\left(e^{-r}r^{a+b}\right)$$

Wow. Aren't you glad you didn't slog through the math? Here are the first few generalized Laguerre polynomials:

- ✔ $L_0^b(r) = 1$

- ✔ $L_1^b(r) = -r + b + 1$

- ✔ $L_2^b(r) = \dfrac{-r^2}{2} - (b+2)r + \dfrac{(b+2)(b+1)}{2}$

- ✔ $L_3^b(r) = \dfrac{-r^3}{6} - \dfrac{(b+3)r^2}{2} - \dfrac{(b+2)(b+3)r}{2} + \dfrac{(b+1)(b+2)(b+3)}{6}$

All right, you have the form for $R_{nl}(r)$. To find the complete wave function, $\psi_{nlm}(r, \theta, \phi)$, you multiply by the spherical harmonics, $Y_{lm}(\theta, \phi)$:

$$\psi_{nlm}(r,\theta,\phi) = R_{nl}(r)Y_{lm}(\theta,\phi)$$

Now take a look at the first few wave functions for the isotropic harmonic oscillator in spherical coordinates:

- ✔ $\psi_{000}(r,\theta,\phi) = \dfrac{2}{\pi^{\frac{1}{4}}}\left(\dfrac{m\omega}{\hbar}\right)^{\frac{3}{4}}\exp\left(-m\omega\dfrac{r^2}{2\hbar}\right)Y_{00}(\theta,\phi)$

- ✔ $\psi_{11m}(r,\theta,\phi) = \dfrac{\left(\dfrac{8}{3}\right)^{\frac{1}{2}}}{\pi^{\frac{1}{4}}}-\left(\dfrac{m\omega}{\hbar}\right)^{\frac{5}{4}}r\,\exp\left(-m\omega\dfrac{r^2}{2\hbar}\right)Y_{1m}(\theta,\phi)$

- ✔ $\psi_{200}(r,\theta,\phi) = \dfrac{\left(\dfrac{8}{3}\right)^{\frac{1}{2}}}{\pi^{\frac{1}{4}}}-\left(\dfrac{m\omega}{\hbar}\right)^{\frac{3}{4}}\left(\dfrac{3}{2}-\dfrac{m\omega r^2}{\hbar}\right)\exp\left(-m\omega\dfrac{r^2}{2\hbar}\right)Y_{00}(\theta,\phi)$

- ✔ $\psi_{31m}(r,\theta,\phi) = \dfrac{\left(\dfrac{16}{15}\right)^{\frac{1}{2}}}{\pi^{\frac{1}{4}}}-\left(\dfrac{m\omega}{\hbar}\right)^{\frac{7}{4}}r^2\,\exp\left(-m\omega\dfrac{r^2}{2\hbar}\right)Y_{1m}(\theta,\phi)$

As you can see, when you have a potential that depends on $r^2$, as with harmonic oscillators, the wave function gets pretty complex pretty fast.

The energy of an isotropic 3D harmonic oscillator is quantized, and you can derive the following relation for the energy levels:

$$E_n = \left(n + \dfrac{3}{2}\right)\hbar\omega \quad n = 1,2,3\ldots$$

So the energy levels start at $3\hbar\omega/2$ and then go to $5\hbar\omega/2$, $7\hbar\omega/2$, and so on.

# Chapter 9

# Understanding Hydrogen Atoms

**N**ot only is hydrogen the most common element in the universe, but it's also the simplest. And one thing quantum physics is good at is predicting everything about simple atoms. This chapter is all about the hydrogen atom and solving the Schrödinger equation to find the energy levels of the hydrogen atom. For such a small little guy, the hydrogen atom can whip up a lot of math — and I solve that math in this chapter.

Using the Schrödinger equation tells you just about all you need to know about the hydrogen atom, and it's all based on a single assumption: that the wave function must go to zero as $r$ goes to infinity, which is what makes solving the Schrödinger equation possible. I start by introducing the Schrödinger equation for the hydrogen atom and take you through calculating energy degeneracy and figuring out how far the electron is from the proton.

## Coming to Terms: The Schrödinger Equation for the Hydrogen Atom

Hydrogen atoms are composed of a single proton, around which rotates a single electron. You can see how that looks in Figure 9-1.

Note that the proton isn't at the exact center of the atom — the center of mass is at the exact center. In fact, the proton is at a radius of $r_p$ from the exact center, and the electron is at a radius of $r_e$.

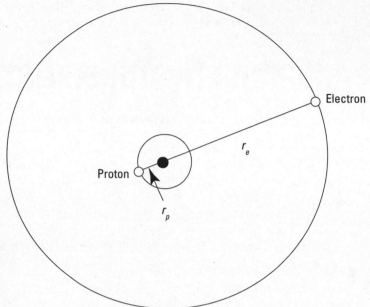

**Figure 9-1:**
The hydro-
gen atom.

So what does the Schrödinger equation, which will give you the wave equations you need, look like? Well, it includes terms for the kinetic and potential energy of the proton and the electron. Here's the term for the proton's kinetic energy:

$$\frac{-\hbar^2}{2m_p}\nabla_p^2$$

where $\nabla_p^2 = \frac{\partial^2}{\partial x_p^2} + \frac{\partial^2}{\partial y_p^2} + \frac{\partial^2}{\partial z_p^2}$ . Here, $x_p$ is the proton's $x$ position, $y_p$ is the proton's $y$ position, and $z_p$ is its $z$ position.

The Schrödinger equation also includes a term for the electron's kinetic energy:

$$\frac{-\hbar^2}{2m_e}\nabla_e^2$$

where $\nabla_e^2 = \frac{\partial^2}{\partial x_e^2} + \frac{\partial^2}{\partial y_e^2} + \frac{\partial^2}{\partial z_e^2}$ . Here, $x_e$ is the electron's $x$ position, $y_e$ is the electron's $y$ position, and $z_e$ is its $z$ position.

Besides the kinetic energy, you have to include the potential energy, V($r$), in the Schrödinger equation, which makes the time-independent Schrödinger equation look like this:

$$\frac{-\hbar^2}{2m_p}\nabla_p^{\,2}\psi\left(r_e,r_p\right)-\frac{\hbar^2}{2m_e}\nabla_e^{\,2}\psi\left(r_e,r_p\right)+V\left(r\right)\psi\left(r_e,r_p\right)=E\psi\left(r_e,r_p\right)$$

where $\psi(r_e,\,r_p)$ is the electron and proton's wave function.

The electrostatic potential energy, V($r$), for a central potential is given by the following formula, where $r$ is the radius vector separating the two charges:

$$V\left(r\right)=-\frac{1}{4\pi\varepsilon_0}\frac{e^2}{|r|}$$

As is common in quantum mechanics, you use CGS (centimeter-gram-second) system of units, where $1=\frac{1}{4\pi\varepsilon_0}$.

So the potential due to the electron and proton charges in the hydrogen atom is

$$V\left(r\right)=\frac{-e^2}{|r|}$$

Note that $r = r_e - r_p$, so the preceding equation becomes

$$V\left(r\right)=\frac{-e^2}{\left|r_e-r_p\right|}$$

which gives you this Schrödinger equation:

$$\frac{-\hbar^2}{2m_p}\nabla_p^{\,2}\psi\left(r_e,r_p\right)-\frac{\hbar^2}{2m_e}\nabla_e^{\,2}\psi\left(r_e,r_p\right)-\frac{e^2}{\left|r_e-r_p\right|}\psi\left(r_e,r_p\right)=E\psi\left(r_e,r_p\right)$$

Okay, so how do you handle this equation? Find out in the next section.

# Simplifying and Splitting the Schrödinger Equation for Hydrogen

Here's the usual quantum mechanical Schrödinger equation for the hydrogen atom:

$$\frac{-\hbar^2}{2m_p}\nabla_p^{\;2}\psi\left(r_e,r_p\right) - \frac{\hbar^2}{2m_e}\nabla_e^{\;2}\psi\left(r_e,r_p\right) - \frac{e^2}{\left|r_e - r_p\right|}\psi\left(r_e,r_p\right) = E\psi\left(r_e,r_p\right)$$

The problem is that you're taking into account the distance the proton is from the center of mass of the atom, so the math is messy. If you were to assume that the proton is stationary and that $r_p = 0$, this equation would break down to the following, which is much easier to solve:

$$\frac{-\hbar^2}{2m_e}\nabla_e^{\;2}\psi\left(r_e\right) - \frac{e^2}{\left|r_e\right|}\psi\left(r_e\right) = E\psi\left(r_e\right)$$

Unfortunately, that equation isn't exact because it ignores the movement of the proton, so you see the more-complete version of the equation in quantum mechanics texts.

To simplify the usual Schrödinger equation, you switch to center-of-mass coordinates. The center of mass of the proton/electron system is at this location:

$$\mathbf{R} = \frac{m_e r_e + m_p r_p}{m_e + m_p}$$

And the vector between the electron and proton is

$$r = r_e - r_p$$

Using vectors $\mathbf{R}$ and $r$ instead of $r_e$ and $r_p$ makes the Schrödinger equation easier to solve. The Laplacian for $\mathbf{R}$ is $\nabla_{\mathbf{R}}^{\;2} = \frac{\partial^2}{\partial X^2} + \frac{\partial^2}{\partial Y^2} + \frac{\partial^2}{\partial Z^2}$. And the Laplacian for $r$ is $\nabla_r^{\;2} = \frac{\partial^2}{\partial x^2} + \frac{\partial^2}{\partial y^2} + \frac{\partial^2}{\partial z^2}$.

How can you relate $\nabla_{\mathbf{R}}^{\;2}$ and $\nabla_r^{\;2}$ to the usual equation's $\nabla_p^{\;2}$ and $\nabla_e^{\;2}$? After the algebra settles, you get

$$\frac{1}{m_e}\nabla_e^{\;2} + \frac{1}{m_p}\nabla_p^{\;2} = \frac{1}{M}\nabla_{\mathbf{R}}^{\;2} + \frac{1}{m}\nabla_r^{\;2}$$

where M = $m_e$ + $m_p$ is the total mass and $m = \dfrac{m_e m_p}{m_e + m_p}$ is called the *reduced mass*. When you put together the equations for the center of mass, the vector between the proton and the electron, the total mass, and $m$, then the time-independent Schrödinger equation becomes the following:

$$\frac{-\hbar^2}{2M}\nabla_R^{\,2}\psi(\mathbf{R},\mathbf{r}) - \frac{\hbar^2}{2m}\nabla_r^{\,2}\psi(\mathbf{R},\mathbf{r}) + V(\mathbf{r})\psi(\mathbf{R},\mathbf{r})$$

From $V(\mathbf{r}) = \dfrac{-e^2}{\left| r_e - r_p \right|}$, $V(\mathbf{r}) = \dfrac{-e^2}{\left| r_e - r_p \right|}$. And because $r = r_e - r_p$, this equation

becomes $V(\mathbf{r}) = \dfrac{-e^2}{\left| r \right|}$. So $\dfrac{-\hbar^2}{2M}\nabla_R^{\,2}\psi(\mathbf{R},\mathbf{r}) - \dfrac{\hbar^2}{2m}\nabla_r^{\,2}\psi(\mathbf{R},\mathbf{r}) + V(\mathbf{r})\psi(\mathbf{R},\mathbf{r})$

$= E\psi(\mathbf{R},\mathbf{r})$ becomes $\dfrac{-\hbar^2}{2M}\nabla_R^{\,2}\psi(\mathbf{R},\mathbf{r}) - \dfrac{\hbar^2}{2m}\nabla_r^{\,2}\psi(\mathbf{R},\mathbf{r}) - \dfrac{e^2}{\left| r \right|}\psi(\mathbf{R},\mathbf{r}) = E\psi(\mathbf{R},\mathbf{r})$

This looks easier — the main improvement being that you now have $|r|$ in the denominator of the potential energy term rather than $|r_e - r_p|$.

Because the equation contains terms involving either **R** or **r** but not both, the form of this equation indicates that it's a separable differential equation. And that means you can look for a solution of the following form:

ψ(**R**, *r*) = ψ(**R**)ψ(*r*)

Substituting the preceding equation into the one before it gives you the following:

$$\frac{-\hbar^2}{2M}\nabla_R^{\,2}\psi(\mathbf{R})\psi(r) - \frac{\hbar^2}{2m}\nabla_r^{\,2}\psi(\mathbf{R})\psi(r) - \frac{e^2}{\left| r \right|}\psi(\mathbf{R})\psi(r) = E\psi(\mathbf{R})\psi(r)$$

And dividing this equation by ψ(**R**)ψ(*r*) gives you

$$\frac{-\hbar^2}{2M\psi(\mathbf{R})}\nabla_R^{\,2}\psi(\mathbf{R}) - \frac{\hbar^2}{2m\psi(r)}\nabla_r^{\,2}\psi(r) - \frac{e^2}{\left| r \right|} = E$$

Well, well, well. This equation has terms that depend on either $\psi(\mathbf{R})$ or $\psi(r)$ but not both. That means you can separate this equation into *two* equations, like this (where the total energy, E, equals $E_R + E_r$):

➤ $\dfrac{-\hbar^2}{2M\psi(\mathbf{R})}\nabla_{\mathbf{R}}^2 \psi(\mathbf{R}) = E_R$

➤ $\dfrac{-\hbar^2}{2m\psi(r)}\nabla_r^2 \psi(r) - \dfrac{e^2}{|r|} = E_r$

Multiplying $\dfrac{-\hbar^2}{2M\psi(\mathbf{R})}\nabla_{\mathbf{R}}^2 \psi(\mathbf{R}) = E_R$ by $\psi(\mathbf{R})$ gives you

$\dfrac{-\hbar^2}{2M}\nabla_{\mathbf{R}}^2 \psi(\mathbf{R}) = E_R \psi(\mathbf{R})$

And multiplying $\dfrac{-\hbar^2}{2m\psi(r)}\nabla_r^2 \psi(r) - \dfrac{e^2}{|r|} = E_r$ by $\psi(r)$ gives you

$\dfrac{-\hbar^2}{2m}\nabla_r^2 \psi(r) - \dfrac{e^2}{|r|}\psi(r) = E_r \psi(r)$

Now you have two Schrödinger equations. The next two sections show you how to solve them independently.

# Solving for ψ (R)

In $\dfrac{-\hbar^2}{2M}\nabla_{\mathbf{R}}^2 \psi(\mathbf{R}) = E_R \psi(\mathbf{R})$, how do you solve for $\psi(R)$, which is the wave function of the center of mass of the electron/proton system? This is a straightforward differential equation, and the solution is

$\psi(\mathbf{R}) = Ce^{-i\mathbf{k}\cdot r}$

Here, C is a constant and $\mathbf{k}$ is the wave vector, where $|\mathbf{k}| = \dfrac{2ME_R^{\frac{1}{2}}}{\hbar^2}$. You can find C by insisting that $\psi(\mathbf{R})$ be normalized, which means that

$1 = \int\limits_0^\infty \psi(\mathbf{R})\psi^*(\mathbf{R})d^3\mathbf{R}$

This equation tells you that $C = \dfrac{1}{(2\pi)^{\frac{3}{2}}}$. Therefore,

$\psi(\mathbf{R}) = \dfrac{e^{-i\mathbf{k}\cdot r}}{(2\pi)^{\frac{3}{2}}}$

In practice, $E_R$ is so small that people almost always just ignore $\psi(\mathbf{R})$ — that is, they assume it to be 1. In other words, the real action is in $\psi(\mathbf{r})$, not in $\psi(\mathbf{R})$; $\psi(\mathbf{R})$ is the wave function for the center of mass of the hydrogen atom, and $\psi(\mathbf{r})$ is the wave function for a (fictitious) particle of mass $m$.

# Solving for $\psi(r)$

The Schrödinger equation for $\psi(\mathbf{r})$ is the wave function for a made-up particle of mass $m$ (in practice, $m \approx m_e$ and $\psi(\mathbf{r})$ is pretty close to $\psi(\mathbf{r}_e)$, so the energy, $E_r$, is pretty close to the electron's energy). Here's the Schrödinger equation for $\psi(\mathbf{r})$:

$$\frac{-\hbar^2}{2m}\nabla_r^2\psi(\mathbf{r}) - \frac{e^2}{|\mathbf{r}|}\psi(\mathbf{r}) = E_r\psi(\mathbf{r})$$

You can break the solution, $\psi(\mathbf{r})$, into a radial part and an angular part (see Chapter 8):

$$\psi(\mathbf{r}) = R_{nl}(\mathbf{r})Y_{lm}(\theta, \phi)$$

The angular part of $\psi(\mathbf{r})$ is made up of spherical harmonics, $Y_{lm}(\theta, \phi)$, so that part's okay. Now you have to solve for the radial part, $R_{nl}(r)$. Here's what the Schrödinger equation becomes for the radial part:

$$\frac{-\hbar^2}{2m}\frac{d^2}{dr^2}\left[r\,R_{nl}(r)\right] + l(l+1)\frac{\hbar^2}{2mr^2}r\,R_{nl}(r) - \frac{e^2}{r}r\,R_{nl}(r) = E_r r\,R_{nl}(r)$$

where $r = |\mathbf{r}|$. To solve this equation, you take a look at two cases — where $r$ is very small and where $r$ is very large. Putting them together gives you the rough form of the solution.

## Solving the radial Schrödinger equation for small r

For small $r$, the radial wave function must vanish, and you have

$$\frac{-\hbar^2}{2m}\frac{d^2}{dr^2}\left[r\,R_{nl}(r)\right] + l(l+1)\frac{\hbar^2}{2mr^2}r\,R_{nl}(r) = 0$$

And multiplying by $2m/\hbar^2$, you get

$$\frac{-d^2}{dr^2}\left[r\,R_{nl}(r)\right] + \frac{l(l+1)}{r^2}r\,R_{nl}(r) = 0$$

The solution to this equation is proportional to

$$R_{nl}(r) \sim Ar^l + Br^{-l-1}$$

Note, however, that $R_{nl}(r)$ must vanish as $r$ goes to zero — but the $r^{-l-1}$ term goes to infinity. And that means that B must be zero, so you have this solution for small $r$:

$$R_{nl}(r) \sim r_l$$

That takes care of small $r$. The next section takes a look at very large $r$.

## Solving the radial Schrödinger equation for large r

For very large $r$, $\dfrac{-\hbar^2}{2m}\nabla_r^2\psi(r) - \dfrac{e^2}{|r|}\psi(r) = E_r\psi(r)$ becomes

$$\frac{d^2}{dr^2}\left[r\,R_{nl}(r)\right] + \frac{2mE}{\hbar^2}r\,R_{nl}(r) = 0$$

Because the electron is in a bound state in the hydrogen atom, E < 0; thus, the solution to the preceding equation is proportional to

$$R_{nl}(r) \sim Ae^{-\lambda r} + Be^{\lambda r}$$

where $\lambda = \dfrac{(-2mE)^{\frac{1}{2}}}{\hbar}$.

Note that $R_{nl}(r) \sim Ae^{-\lambda r} + Be^{\lambda r}$ diverges as $r$ goes to infinity because of the $Be^{\lambda r}$ term, so B must be equal to zero. That means that $R_{nl}(r) \sim e^{-\lambda r}$. In the next section, you put the solutions for small $r$ and large $r$ together.

## You got the power: Putting together the solution for the radial equation

Putting together the solutions for small $r$ and large $r$ (see the preceding sections), the Schrödinger equation gives you a solution to the radial Schrödinger equation of $R_{nl}(r) = r^l f(r)e^{-\lambda r}$, where f($r$) is some as-yet-undetermined function of $r$. Your next task is to determine f($r$), which you do by substituting this equation into the radial Schrödinger equation, giving you the following:

$$\frac{-\hbar^2}{2m}\frac{d^2}{dr^2}\left[r\,\mathrm{R}_{nl}(r)\right]+l(l+1)\frac{\hbar^2}{2mr^2}r\,\mathrm{R}_{nl}(r)-\frac{e^2}{r}r\,\mathrm{R}_{nl}(r)=\mathrm{E}_r r\,\mathrm{R}_{nl}(r)$$

Performing the substitution gives you the following differential equation:

$$\frac{d^2}{dr^2}\mathrm{f}(r)+2\left[\frac{l(l+1)}{r}-\lambda\right]\frac{d\mathrm{f}(r)}{dr}+2\left[\frac{\frac{me^2}{\hbar^2}-\lambda(l+1)}{r}\right]\mathrm{f}(r)=0$$

Quite a differential equation, eh? But just sit back and relax — you solve it with a power series, which is a common way of solving differential equations. Here's the power-series form of f($r$) to use:

$$\mathrm{f}(r)=\sum_{k=0}^{\infty}a_k r^k$$

Substituting the preceding equation into the one before it gives you

$$\sum_{k=0}^{\infty}\left[k(k+2l+1)a_k r^{k-2}+2\left(\frac{me^2}{\hbar^2}-\lambda(k+l+1)\right)a_k r^{k-1}\right]=0$$

Changing the index of the second term from $k$ to $k-1$ gives you

$$\sum_{k=0}^{\infty}\left[k(k+2l+1)a_k r^{k-2}+2\left(\frac{me^2}{\hbar^2}-\lambda(k+l)\right)a_{k-1}r^{k-2}\right]=0$$

Because each term in this series has to be zero, you have

$$k(k+2l+1)a_k r^{k-2}=2\left[\lambda(k+l)-\frac{me^2}{\hbar^2}\right]a_{k-1}r^{k-2}$$

Dividing by $r^{k-2}$ gives you

$$k(k+2l+1)a_k=2\left[\lambda(k+l)-\frac{me^2}{\hbar^2}\right]a_{k-1}$$

This equation gives the recurrence relation of the infinite series,

$$\sum_{k=0}^{\infty} \left[ k(k+2l+1)a_k r^{k-2} + 2\left( \frac{me^2}{\hbar^2} - \lambda(k+l+1) \right)a_k r^{k-1} \right] = 0.$$ That is, if you have one coefficient, you can get the next one using this equation. What does that buy you? Well, take a look at the ratio of $a_k/a_{k-1}$:

$$\frac{a_k}{a_{k-1}} = \frac{2\left[ \lambda(k+1) - \frac{me^2}{\hbar^2} \right]}{k(k+2l+1)}$$

Here's what this ratio approaches as $k$ goes to $\infty$:

$$\lim_{k \to \infty} \frac{a_k}{a_{k-1}} \to \frac{2\lambda}{k}$$

This resembles the expansion for $e^x$, which is

$$e^x = \sum_{k=0}^{\infty} \left[ \frac{(2x)^k}{k!} \right] = 0$$

As for $e^x$, the ratio of successive terms is

$$\frac{(2x)^k}{k!} \frac{(k-1)!}{(2x)^{k-1}}$$

And in the limit $k \to \infty$, the expansion for $e^x$ approaches $\frac{2}{k}$:

$$\frac{(2x)^k}{k!} \frac{(k-1)!}{(2x)^{k-1}} \to \frac{2}{k} \qquad k \to \infty$$

That's the case for $e^x$. For f($r$), you have

$$\lim_{k \to \infty} \frac{a_k}{a_{k-1}} \to \frac{2\lambda}{k}$$

Comparing these two equations, it's apparent that

$$f(r) = \sum_{k=0}^{\infty} a_k r^k = e^{2\lambda r}$$

The radial wave function, $R_{nl}(r)$, looks like this:

$$R_{nl}(r) = r^l f(r) e^{-\lambda r}$$

where $\lambda = \dfrac{(-2mE)^{\frac{1}{2}}}{\hbar}$ .

Plugging the form you have for $f(r)$, $f(r) = \sum\limits_{k=0}^{\infty} a_k r^k = e^{2\lambda r}$, into $R_{nl}(r) = r^l f(r) e^{-\lambda r}$ gives you the following:

$$R_{nl}(r) = r^l e^{2\lambda r} e^{-\lambda r}$$
$$= r^l e^{\lambda r}$$

Okay, should you be overjoyed? Well, no. Here's what the wave function $\psi(r)$ looks like: $\psi(r) = R_{nl}(r)\, Y_{lm}(\theta, \phi)$. And substituting in your form of $R_{nl}(r)$ from this equation gives you

$$\psi(r) = r^l e^{\lambda r}\, Y_{lm}(\theta, \phi)$$

That looks fine — except that it goes to infinity as $r$ goes to infinity. You expect $\psi(r)$ to go to zero as $r$ goes to infinity, so this version of $R_{nl}(r) = r^l e^{\lambda r}$ is clearly unphysical. In other words, something went wrong somewhere. How can you fix this version of $f(r)$?

## Fixing f (r) to keep it finite

You need the solution for the radial equation to go to zero as $r$ goes to infinity. The problem of having $\psi(r)$ go to infinity as $r$ goes to infinity lies in the form you assume for $f(r)$ in the preceding section, which is

$$f(r) = \sum_{k=0}^{\infty} a_k r^k$$

The solution is to say that this series must terminate at a certain index, which you call N. N is called the *radial quantum number*. So this equation becomes the following (note that the summation is now to N, not infinity):

$$f(r) = \sum_{k=0}^{N} a_k r^k$$

For this series to terminate, $a_{N+1}$, $a_{N+2}$, $a_{N+3}$, and so on must all be zero. The recurrence relation for the coefficients $a_k$ is

$$k(k+2l+1)a_k = 2\left[\lambda(k+l)-\frac{me^2}{\hbar^2}\right]a_{k-1}$$

For $a_{N+1}$ to be zero, the factor multiplying $a_{k-1}$ must be zero for $k = N + 1$, which means that

$$2\left[\lambda(k+l)-\frac{me^2}{\hbar^2}\right]=0$$

Substituting in $k = N + 1$ gives you $2\left[\lambda(N+l+1)-\frac{me^2}{\hbar^2}\right]=0$. And dividing by 2 gives you $\lambda(N+l+1)-\frac{me^2}{\hbar^2}=0$. Making the substitution $N + l + 1 \rightarrow n$, where $n$ is called the *principal quantum number,* gives you

$$n\lambda-\frac{me^2}{\hbar^2}=0 \quad n=1,2,3...$$

This is the quantization condition that must be met if the series for f($r$) is to be finite, which it must be, physically:

$$f(r)=\sum_{k=0}^{N}a_k r^k$$

Because $\lambda=\frac{(-2mE)^{\frac{1}{2}}}{\hbar}$, the equation $n\lambda-\frac{me^2}{\hbar^2}=0$ puts constraints on the allowable values of the energy.

# Finding the allowed energies of the hydrogen atom

The quantization condition for $\psi(r)$ to remain finite as $r$ goes to infinity is

$$n\lambda-\frac{me^2}{\hbar^2}=0 \qquad n=1,2,3...$$

where $\lambda=\frac{(-2mE)^{\frac{1}{2}}}{\hbar}$. Substituting $\lambda$ into the quantization-condition equation gives you the following:

$$\frac{n(-2m\mathrm{E})^{\frac{1}{2}}}{\hbar} - \frac{me^2}{\hbar^2} = 0 \qquad n = 1,2,3\dots$$

$$\frac{n(-2m\mathrm{E})^{\frac{1}{2}}}{\hbar} = \frac{me^2}{\hbar^2}$$

Now solve for the energy, E. Squaring both sides of the preceding equation gives you

$$n^2 \frac{(-2m\mathrm{E})}{\hbar^2} = \frac{m^2 e^4}{\hbar^4} \qquad n = 1,2,3\dots$$

$$-n^2 2\mathrm{E} = \frac{me^4}{\hbar^2}$$

So here's the energy, E (***Note:*** Because E depends on the principal quantum number, I've renamed it $E_n$):

$$\mathrm{E}_n = \frac{-me^4}{2n^2\hbar^2} \qquad n = 1,2,3\dots$$

Physicists often write this result in terms of the *Bohr radius* — the orbital radius that Niels Bohr calculated for the electron in a hydrogen atom, $r_0$. The Bohr radius is $r_0 = \dfrac{\hbar^2}{me^2}$ .

And in terms of $r_0$, here's what $E_n$ equals:

$$\mathrm{E}_n = \frac{-me^4}{2n^2\hbar^2} = \frac{-e^2}{2r_0}\frac{1}{n^2} \qquad n = 1,2,3\dots$$

The ground state, where $n = 1$, works out to be about E = –13.6 eV.

Notice that this energy is negative because the electron is in a bound state — you'd have to add energy to the electron to free it from the hydrogen atom. Here are the first and second excited states:

> ✔ **First excited state, $n = 2$:** E = –3.4 eV
>
> ✔ **Second excited state, $n = 3$:** E = –1.5 eV

Okay, now you've used the quantization condition, which is

$$n\lambda - \frac{me^2}{\hbar^2} = 0 \qquad n = 1,2,3\dots$$

to determine the energy levels of the hydrogen atom.

# Getting the form of the radial solution of the Schrödinger equation

In this section, you complete the calculation of the wave functions. Go to the calculation of $R_{nl}(r)$ (see the earlier section titled "You got the power: Putting together the solution for the radial equation"). So far, you know that

$$R_{nl}(r) = r^l f(r) e^{-\lambda r}, \text{ where } f(r) = \sum_{k=0}^{N} a_k r^k. \text{ Therefore,}$$

$$R_{nl}(r) = r^l e^{-\lambda r} \sum_{k=0}^{N} a_k r^k$$

In fact, this isn't quite enough; the preceding equation comes from solving the radial Schrödinger equation:

$$\frac{-\hbar^2}{2m} \frac{d^2}{dr^2} r\, R_{nl}(r) + l(l+1)\frac{\hbar^2}{2mr^2} r\, R_{nl}(r) - \frac{e^2}{r} r\, R_{nl}(r) = E_r r\, R_{nl}(r)$$

The solution is only good to a multiplicative constant, so you add such a constant, $A_{nl}$ (which turns out to depend on the principal quantum number $n$ and the angular momentum quantum number $l$), like this:

$$R_{nl}(r) = A_{nl} r^l e^{-\lambda r} \sum_{k=0}^{N} a_k r^k$$

You find $A_{nl}$ by normalizing $R_{nl}(r)$.

Now try to solve for $R_{nl}(r)$ by just flat-out doing the math. For example, try to find $R_{10}(r)$. In this case, $n = 1$ and $l = 0$. Then, because $N + l + 1 = n$, you have $N = n - l - 1$. So $N = 0$ here. That makes $R_{nl}(r)$ look like this:

$$R_{10}(r) = A_{nl} r^l e^{-\lambda r} \sum_{k=0}^{0} a_k r^k$$

And the summation in this equation is equal to $\sum_{k=0}^{0} a_k r^k = a_0$, so

$$R_{10}(r) = A_{10} r^l e^{-\lambda r} a_0$$

And because $l = 0$, $r^l = 1$, so $R_{10}(r) = A_{10} e^{-\lambda r} a_0$, where $\lambda = \frac{(-2mE)^{\frac{1}{2}}}{\hbar}$. Therefore, you can also write $R_{10}(r) = A_{10} e^{-\lambda r} a_0$ as

$$R_{10}(r) = A_{10} \exp\left(\frac{-r}{nr_0}\right) a_0$$

where $r_0$ is the Bohr radius. To find $A_{10}$ and $a_0$, you normalize $\psi_{100}(r,\ \theta,\ \phi)$ to 1, which means integrating $|\psi_{100}(r,\ \theta,\ \phi)|^2 d^3 r$ over all space and setting the result to 1.

Now $d^3 r = r^2 \sin\theta\ dr\ d\theta\ d\phi$, and integrating the spherical harmonics, such as $Y_{00}$, over a complete sphere, $\int |Y_{00}|^2 \sin\theta d\theta d\phi$, gives you 1. Therefore, you're left with the radial part to normalize:

$$1 = \int_0^{+\infty} r^2 \left| R_{10}(r) \right|^2 dr$$

Plugging $R_{10}(r) = A_{10}\exp\left(\dfrac{-r}{nr_0}\right)a_0$ into $1 = \int_0^{+\infty} r^2 \left| R_{10}(r) \right|^2 dr$ gives you

$$1 = A_{10}{}^2 a_0{}^2 \int_0^{+\infty} r^2 \exp\left(\dfrac{-2r}{nr_0}\right) dr$$

You can solve this kind of integral with the following relation:

$$\int_0^{+\infty} x^n \exp(ax)\,dx = \dfrac{n!}{a^{n+1}}$$

With this relation, the equation $1 = A_{10}{}^2 a_0{}^2 \int_0^{+\infty} r^2 \exp\left(\dfrac{-2r}{nr_0}\right) dr$ becomes

$$1 = A_{10}{}^2 a_0{}^2 \int_0^{+\infty} r^2 \exp\left(\dfrac{-2r}{nr_0}\right) dr = A_{10}{}^2 a_0{}^2 \dfrac{r_0{}^3}{4}$$

Therefore,

$$A_{10}{}^2 a_0{}^2 \dfrac{r_0{}^3}{4} = 1$$

$$A_{10}{}^2 a_0{}^2 = \dfrac{4}{r_0{}^3}$$

$$A_{10} a_0 = \dfrac{2}{r_0{}^{\frac{3}{2}}}$$

This is a fairly simple result. Because $A_{10}$ is just there to normalize the result, you can set $A_{10}$ to 1 (this wouldn't be the case if $A_{10}{}^2 a_0{}^2 \dfrac{r_0{}^3}{4} = 1$ involved multiple terms). Therefore, $a_0 = \dfrac{2}{r_0{}^{\frac{3}{2}}}$. That's fine, and it makes $R_{10}(r)$, which is

$$R_{10}(r) = A_{10}\exp\left(\dfrac{-r}{nr_0}\right)a_0$$

$$R_{10}(r) = \dfrac{2}{r_0{}^{\frac{3}{2}}} e^{\frac{-r}{r_0}}$$

You know that $\psi_{nlm}(r, \theta, \phi) = R_{nl}(r)\, Y_{lm}(\theta, \phi)$.

And so $\psi_{100}(r, \theta, \phi)$ becomes

$$\psi_{100}(r,\theta,\phi) = \frac{2}{r_0^{\frac{3}{2}}} e^{\frac{-r}{r_0}} Y_{00}(\theta,\phi)$$

Whew. In general, here's what the wave function $\psi_{nlm}(r, \theta, \phi)$ looks like for hydrogen:

$$\psi_{nlm}(r,\theta,\phi) = \frac{\left(\dfrac{2}{nr_0}\right)^{\frac{3}{2}} \left[(n-l-1)!\right]^{\frac{1}{2}}}{\left[2n(n+1)!\right]^{\frac{1}{2}}} e^{\frac{-r}{nr_0}} \left(\frac{2r}{nr_0}\right)^{l} L_{n-l-1}^{2l+1}\left(\frac{2r}{nr_0}\right) Y_{lm}(\theta,\phi)$$

where $L_{n-l-1}^{2l+1}(2r/nr_0)$ is a generalized Laguerre polynomial. Here are the first few generalized Laguerre polynomials:

- $L_0^{b}(r) = 1$
- $L_1^{b}(r) = -r + b + 1$
- $L_2^{b}(r) = \dfrac{-r^2}{2} - (b+2)r + \dfrac{(b+2)(b+1)}{2}$
- $L_3^{b}(r) = \dfrac{-r^3}{6} - \dfrac{(b+3)r^2}{2} + \dfrac{(b+2)(b+3)r}{2} + \dfrac{(b+1)(b+2)(b+3)}{6}$

## Some hydrogen wave functions

So what do the hydrogen wave functions look like? In the preceding section, you find that $\psi_{100}(r, \theta, \phi)$ looks like this:

$$\psi_{100}(r,\theta,\phi) = \frac{2}{r_0^{\frac{3}{2}}} e^{\frac{-r}{r}} Y_{00}(\theta,\phi)$$

Here are some other hydrogen wave functions:

- $\psi_{200}(r,\theta,\phi) = \dfrac{1}{2^{\frac{1}{2}} r_0^{\frac{3}{2}}} e^{\frac{-r}{2r}} Y_{00}(\theta,\phi)$

- $\psi_{21m}(r,\theta,\phi) = \dfrac{1}{6^{\frac{1}{2}} r_0^{\frac{3}{2}}} \dfrac{r}{2r_0} e^{\frac{-r}{2r}} Y_{1m}(\theta,\phi)$

$$\checkmark \quad \psi_{300}(r,\theta,\phi) = \frac{2}{3\left(3^{\frac{1}{2}}\right)r_0^{\frac{3}{2}}} e^{\frac{-r}{3r_0}} \left[1 - \left(\frac{2r}{3r_0}\right) + \left(\frac{2r^2}{27r_0^2}\right)\right] Y_{00}(\theta,\phi)$$

$$\checkmark \quad \psi_{31m}(r,\theta,\phi) = \frac{8}{9\left(6^{\frac{1}{2}}\right)r_0^{\frac{3}{2}}} e^{\frac{-r}{3r_0}} \frac{r}{3r_0} \left[1 - \left(\frac{r}{6r_0}\right)\right] Y_{1m}(\theta,\phi)$$

$$\checkmark \quad \psi_{32m}(r,\theta,\phi) = \frac{4}{9(30)^{\frac{1}{2}}r_0^{\frac{3}{2}}} \frac{r^2}{9r_0^2} e^{\frac{-r}{3r_0}} Y_{2m}(\theta,\phi)$$

Note that $\psi_{nlm}(r, \theta, \phi)$ behaves like $r^l$ for small $r$ and therefore goes to zero. And for large $r$, $\psi_{nlm}(r, \theta, \phi)$ decays exponentially to zero. So you've solved the problem you had earlier of the wave function diverging as $r$ becomes large — and all because of the quantization condition, which cut the expression for f($r$) from an exponent to a polynomial of limited order. Not bad.

You can see the radial wave function $R_{10}(r)$ in Figure 9-2. $R_{20}(r)$ appears in Figure 9-3. And you can see $R_{21}(r)$ in Figure 9-4.

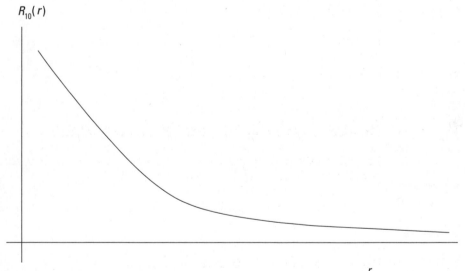

**Figure 9-2:**
The radial wave function $R_{10}(r)$.

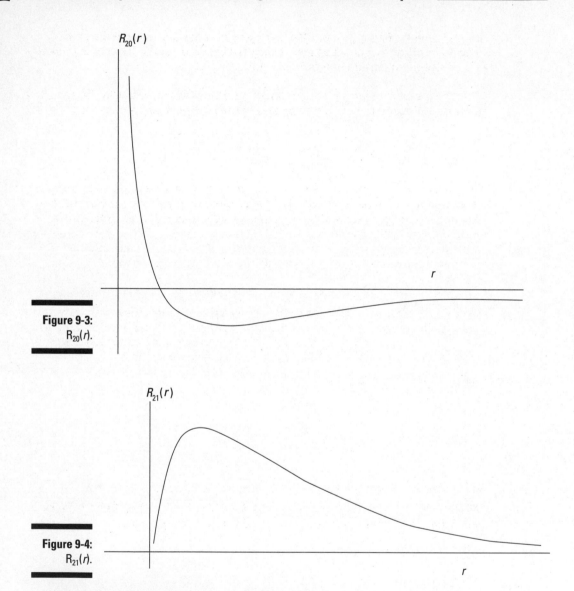

**Figure 9-3:**
$R_{20}(r)$.

**Figure 9-4:**
$R_{21}(r)$.

# Calculating the Energy Degeneracy of the Hydrogen Atom

Each quantum state of the hydrogen atom is specified with three quantum numbers: $n$ (the principal quantum number), $l$ (the angular momentum quantum number of the electron), and $m$ (the $z$ component of the electron's angular momentum, $\psi_{nlm}[r,\,\theta,\,\phi]$). How many of these states have the same energy? In other words, what's the energy degeneracy of the hydrogen atom in terms of the quantum numbers $n$, $l$, and $m$?

Well, the actual energy is just dependent on $n$, as you see earlier in the section titled "Finding the allowed energies of the hydrogen atom":

$$E_n = \frac{-me^4}{2n^2\hbar^2} \quad n = 1, 2, 3\ldots$$

That means the E is independent of $l$ and $m$. So how many states, $|n,\,l,\,m>$, have the same energy for a particular value of $n$? Well, for a particular value of $n$, $l$ can range from zero to $n-1$. And each $l$ can have different values of $m$, so the total degeneracy is

$$\text{Degeneracy} = \sum_{l=0}^{n-1}\left(\text{Degeneracy in } m\right)$$

The degeneracy in $m$ is the number of states with different values of $m$ that have the same value of $l$. For any particular value of $l$, you can have $m$ values of $-l$, $-l+1$, ..., 0, ..., $l-1$, $l$. And that's $(2l+1)$ possible $m$ states for a particular value of $l$. So you can plug in $(2l+1)$ for the degeneracy in $m$:

$$\text{Degeneracy} = \sum_{l=0}^{n-1}\left(2l+1\right)$$

And this series works out to be just $n^2$.

So the degeneracy of the energy levels of the hydrogen atom is $n^2$. For example, the ground state, $n=1$, has *degeneracy* = $n^2$ = 1 (which makes sense because $l$, and therefore $m$, can only equal zero for this state).

For $n=2$, you have a degeneracy of 4:

✔ $\psi_{200}(r,\,\theta,\,\phi)$

✔ $\psi_{21-1}(r,\,\theta,\,\phi)$

✔ $\psi_{210}(r,\,\theta,\,\phi)$

✔ $\psi_{211}(r,\,\theta,\,\phi)$

Cool.

## *Quantum states: Adding a little spin*

You may be asking yourself — what about the spin of the electron? Right you are! The spin of the electron does provide additional quantum states. Up to now in this section, you've been treating the wave function of the hydrogen atom as a product of radial and angular parts:

$$\psi_{nlm}(r,\,\theta,\,\phi) = R_{nl}(r)Y_{lm}(\theta,\,\phi)$$

Now you can add a spin part, corresponding to the spin of the electron, where $s$ is the spin of the electron and $m_s$ is the $z$ component of the spin:

$$\left|s,m_s\right\rangle$$

The spin part of the equation can take the following values:

- $|^1/_2,\,^1/_2>$
- $|^1/_2,\,-^1/_2>$

Hence, $\psi_{nlm}(r,\,\theta,\,\phi)$ now becomes $\psi_{nlmm_s}(r,\,\theta,\,\phi)$:

$$\psi_{nlmm_s}(r,\theta,\phi) = R_{nl}(r)Y_{lm}(\theta,\phi)\left|s,m_s\right\rangle$$

And this wave function can take two different forms, depending on $m_s$, like this:

- $\psi_{nlm\frac{1}{2}}(r,s,\theta,\phi) = R_{nl}(r)Y_{lm}(\theta,\phi)\left|\frac{1}{2},\frac{1}{2}\right\rangle$
- $\psi_{nlm-\frac{1}{2}}(r,s,\theta,\phi) = R_{nl}(r)Y_{lm}(\theta,\phi)\left|\frac{1}{2},-\frac{1}{2}\right\rangle$

In fact, you can use the spin notation (which you use in Chapter 6), where

$$\left|\frac{1}{2},\frac{1}{2}\right\rangle = \left|\begin{matrix}1\\0\end{matrix}\right|$$

$$\left|\frac{1}{2},-\frac{1}{2}\right\rangle = \left|\begin{matrix}0\\1\end{matrix}\right|$$

For example, for $|^1/_2, ^1/_2>$, you can write the wave function as

$$\psi_{nlm\frac{1}{2}}(r,\theta,\phi) = R_{nl}(r)Y_{lm}(\theta,\phi)\begin{vmatrix}1\\0\end{vmatrix}$$

$$= \begin{vmatrix}\psi_{nlm}(r,\theta,\phi)\\0\end{vmatrix}$$

And for $|^1/_2, -^1/_2>$, you can write the wave function as

$$\psi_{nlm-\frac{1}{2}}(r,\theta,\phi) = R_{nl}(r)Y_{lm}(\theta,\phi)\begin{vmatrix}0\\1\end{vmatrix}$$

$$= \begin{vmatrix}0\\\psi_{nlm}(r,\theta,\phi)\end{vmatrix}$$

What does this do to the energy degeneracy? If you include the spin of the electron, there are two spin states for every state $|n, l, m>$, so the degeneracy becomes

$$\text{Degeneracy} = \sum_{l=0}^{n-1}2(2l+1) = 2n^2$$

So if you include the electron's spin, the energy degeneracy of the hydrogen atom is $2n^2$.

In fact, you can even add the spin of the proton to the wave function (although people don't usually do that, because the proton's spin interacts only weakly with magnetic fields applied to the hydrogen atom). In that case, you have a wave function that looks like the following:

$$\psi_{nlmm_{se}m_{sp}}(r,\theta,\phi) = R_{nl}(r)Y_{lm}(\theta,\phi)|s_e,m_{se}\rangle|s_p,m_{sp}\rangle$$

where $s_e$ is the spin of the electron, $m_{se}$ is the $z$ component of the electron's spin, $s_p$ is the spin of the proton, and $m_{sp}$ is the $z$ component of the proton's spin.

If you include the proton's spin, the wave function can now take four different forms, depending on $m_s$, like this:

- $\psi_{nlm\frac{1}{2}\frac{1}{2}}(r,\theta,\phi) = R_{nl}(r)Y_{lm}(\theta,\phi)\left|\frac{1}{2},\frac{1}{2}\right\rangle\left|\frac{1}{2},\frac{1}{2}\right\rangle$

- $\psi_{nlm\frac{1}{2}-\frac{1}{2}}(r,\theta,\phi) = R_{nl}(r)Y_{lm}(\theta,\phi)\left|\frac{1}{2},\frac{1}{2}\right\rangle\left|\frac{1}{2},-\frac{1}{2}\right\rangle$

- $\psi_{nlm-\frac{1}{2}\frac{1}{2}}(r,\theta,\phi) = R_{nl}(r)Y_{lm}(\theta,\phi)\left|\frac{1}{2},-\frac{1}{2}\right\rangle\left|\frac{1}{2},\frac{1}{2}\right\rangle$

- $\psi_{nlm-\frac{1}{2}-\frac{1}{2}}(r,\theta,\phi) = R_{nl}(r)Y_{lm}(\theta,\phi)\left|\frac{1}{2},-\frac{1}{2}\right\rangle\left|\frac{1}{2},-\frac{1}{2}\right\rangle$

The degeneracy must now include the proton's spin, so that's a factor of four for each |*n, l, m*>:

$$\text{Degeneracy} = \sum_{l=0}^{n-1} 4(2l+1)$$

$$= 4n^2$$

# On the lines: Getting the orbitals

When you study heated hydrogen in spectroscopy, you get a spectrum consisting of various lines, named the *s* (for *sharp*), *p* (for *principal*), *d* (for *diffuse*), and *f* (for *fundamental*) lines. And other, unnamed lines are present as well — the *g, h,* and so on.

The *s, p, d, f,* and the rest of the lines turn out to correspond to different angular momentum states of the electron, called *orbitals*. The *s* state corresponds to *l* = 0; the *p* state, to *l* = 1; the *d* state, to *l* = 2; the *f* state, to *l* = 3; and so on. Each of these angular momentum states has a differently shaped electron cloud around the proton — that is, a different orbital.

Three quantum numbers — *n, l,* and *m* — determine orbitals. For example, the electron cloud for the |1, 0, 0> state (1*s*, with *m* = 0) appears in Figure 9-5.

**Figure 9-5:**
The |1, 0, 0>
state.

The |4, 3, 2> state (4*f,* with *m* = 2) appears in Figure 9-6.

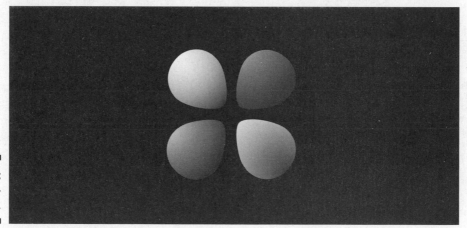

**Figure 9-6:**
The |4, 3, 2>
state.

The $|2, 1, 1>$ state ($2p$, with $m = 1$) appears in Figure 9-7.

# Hunting the Elusive Electron

Just where is the electron at any one time? In other words, how far is the electron from the proton? You can find the expectation value of $r$, that is, $<r>$, to tell you. If the wave function is $\psi_{nlm}(r, \theta, \phi)$, then the following expression represents the probability that the electron will be found in the spatial element $d^3r$:

$$|\psi_{nlm}(r, \theta, \phi)|^2 d^3r$$

In spherical coordinates, $d^3r = r^2 \sin\theta \, dr \, d\theta \, d\phi$. So you can write $|\psi_{nlm}(r, \theta, \phi)|^2 d^3r$ as

$$|\psi_{nlm}(r, \theta, \phi)|^2 r^2 \sin\theta \, dr \, d\theta \, d\phi$$

The probability that the electron is in a spherical shell of radius $r$ to $r + dr$ is therefore

$$\int_0^\pi |\psi_{nlm}(r,\theta,\phi)|^2 \sin\theta \, d\theta \int_0^{2\pi} d\phi \, r^2 \, dr$$

And because $\psi_{nlm}(r, \theta, \phi) = R_{nl}(r)Y_{lm}(\theta, \phi)$, this equation becomes the following:

$$\int_0^\pi \left| R_{nl}(r)Y_{lm}(\theta,\phi)\right|^2 \sin\theta \, d\theta \int_0^{2\pi} d\phi \, r^2 dr$$

The preceding equation is equal to

$$\left| R_{nl}(r)\right|^2 r^2 dr \int_0^\pi \left| Y_{lm}(\theta,\phi)\right|^2 \sin\theta \, d\theta \int_0^{2\pi} d\phi$$

$$\text{or } \left| R_{nl}(r)\right|^2 r^2 dr \int_0^\pi Y_{lm}^*(\theta,\phi)Y_{lm}(\theta,\phi)\sin\theta \, d\theta \int_0^{2\pi} d\phi$$

Spherical harmonics are normalized, so this just becomes

$$|R_{nl}(r)|^2 r^2 \, dr$$

Okay, that's the probability that the electron is inside the spherical shell from $r$ to $r + dr$. So the expectation value of $r$, which is <r>, is

$$\langle r \rangle = \int_0^\infty r \left| R_{nl}(r)\right|^2 dr$$

which is

$$\langle r \rangle = \int_0^\infty r^3 \left| R_{nl}(r)\right|^2 r^2 dr$$

This is where things get more complex, because $R_{nl}(r)$ involves the Laguerre polynomials. But after a lot of math, here's what you get:

$$\langle r \rangle = \int_0^\infty r^3 \left| R_{nl}(r)\right|^2 dr = \left[ 3n^2 - l(l+1)\right]\frac{r_0}{2}$$

where $r_0$ is the Bohr radius: $r_0 = \dfrac{\hbar^2}{me^2}$. The Bohr radius is about $5.29 \times 10^{-11}$ meters, so the expectation value of the electron's distance from the proton is

$$<r> = [3n^2 - l(l+1)](2.65 \times 10^{-11}) \text{ meters}$$

So, for example, in the 1s state ($|1, 0, 0>$), the expectation value of $r$ is equal to

$$<r>_{1s} = 3(2.65 \times 10^{-11}) = 7.95 \times 10^{-11} \text{ meters}$$

And in the 4p state ($| 4, 1, m>$),

$$<r>_{4p} = 46(2.65 \times 10^{-11}) = 1.22 \times 10^{-9} \text{ meters}$$

And that concludes this chapter, which has been a triumph for the Schrödinger equation.

# Chapter 10

# Handling Many Identical Particles

*H*ydrogen atoms (see Chapter 9) involve only a proton and an electron, but all other atoms involve more electrons than that. So how do you deal with multiple-electron atoms? For that matter, how do you deal with multi-particle systems, such as even a simple gas?

In general, you *can't* deal with problems like this — exactly, anyway. Imagine the complexity of just two electrons moving in a helium atom — you'd have to take into account the interaction of the electrons not only with the nucleus of the atom but also with each other — and that depends on their relative positions. So not only does the Hamiltonian have a term in $1/r_1$ for the potential energy of the first electron and $1/r_2$ for the second electron, but it also has

a term in $-\dfrac{1}{|r_1 - r_2|}$ for the potential energy that comes from the interaction of

the two electrons. And that makes an exact wave function just about impossible to find.

However, even without finding exact wave functions, you can still do a surprising amount with multi-particle systems, such as deriving the *Pauli exclusion principle* — which says, among other things, that no two electrons can be in the exact same quantum state. In fact, you'll probably be surprised at how much you can actually say about multi-particle systems using quantum mechanics. This chapter starts with an introduction to many-particle systems and goes on to discuss identical particles, symmetry (and anti-symmetry), and electron shells.

# Many-Particle Systems, Generally Speaking

You can see a multi-particle system in Figure 10-1, where a number of particles are identified by their position (ignore spin for the moment). This section explains how to describe that system in quantum physics terms.

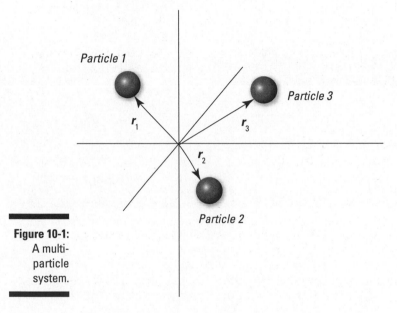

**Figure 10-1:**
A multi-particle system.

# Considering wave functions and Hamiltonians

Begin by working with the wave function. The state of a system with many particles, as shown in Figure 10-1, is given by $\psi(r_1, r_2, r_3, ...)$. And here's the probability that particle 1 is in $d^3r_1$, particle 2 is in $d^3r_2$, particle 3 is in $d^3r_3$, and so on:

$$\left|\psi(r_1, r_2, r_3, ...)\right|^2 d^3r_1 d^3r_2 d^3r_3 ...$$

The normalization of $\psi(r_1, r_2, r_3, ...)$ demands that

$$\int_{-\infty}^{+\infty} \left|\psi(r_1, r_2, r_3, ...)\right|^2 d^3r_1 d^3r_2 d^3r_3 ... = 1$$

Okay, so what about the Hamiltonian, which gives you the energy states? That is, what is H, where $H\psi(r_1, r_2, r_3, ...) = E\psi(r_1, r_2, r_3, ...)$? When you're dealing with a single particle, you can write this as

$$\frac{p^2}{2m}\psi(r) + V(r)\psi(r) = E\psi(r)$$

And you can write this in turn as

$$\frac{p_i^2}{2m}\psi(r_1,r_2,r_3,...) + V(r_i)\psi(r_1,r_2,r_3,...) = \frac{-\hbar^2}{2m}\nabla^2(r_1,r_2,r_3,...) + V(r_i)(r_1,r_2,r_3,...)$$

The total energy of the system is the sum of the energy of all the particles (omitting spin for the moment), so here's how you can generalize the Hamiltonian for multi-particle systems:

$$H\psi(r_1,r_2,r_3,...) = \sum_{i=1}^{N}\frac{p_i^2}{2m_i}\psi(r_1,r_2,r_3,...) + V(r_1,r_2,r_3,...)(r_1,r_2,r_3,...)$$

This, in turn, equals the following:

$$H\psi(r_1,r_2,r_3,...) = \sum_{i=1}^{N}\frac{p_i^2}{2m_i}\psi(r_1,r_2,r_3,...) + V(r_1,r_2,r_3,...)\psi(r_1,r_2,r_3,...)$$

$$= \sum_{i=1}^{N}\frac{-\hbar^2}{2m_i}\nabla_i^2\psi(r_1,r_2,r_3,...) + V(r_1,r_2,r_3,...)\psi(r_1,r_2,r_3,...)$$

Here, $m_i$ is the mass of the $i$th particle and V is the multi-particle potential.

## A Nobel opportunity: Considering multi-electron atoms

This section takes a look at how the Hamiltonian wave function (see the preceding section) would work for a neutral, multi-electron atom. A multi-electron atom, which you see in Figure 10-2, is the most common multi-particle system that quantum physics considers. Here, **R** is the coordinate of the nucleus (relative to the center of mass), $r_1$ is the coordinate of the first electron (relative to the center of mass), $r_2$ the coordinate of the second electron, and so on.

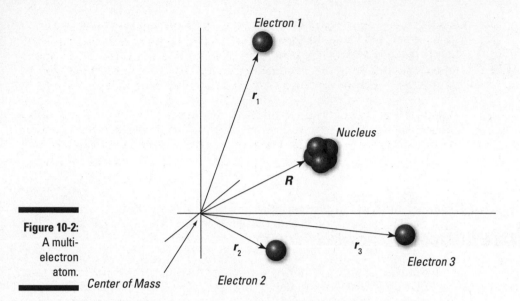

**Figure 10-2:**
A multi-
electron
atom.

If you have Z electrons, the wave function looks like $\psi(r_1, r_2, ..., r_Z, R)$. And the kinetic energy of the electrons and the nucleus looks like this:

$$KE = \sum_{i=1}^{Z} \frac{-\hbar^2}{2m_i} \nabla_i^2 \psi(r_1, r_2, ... r_Z, R) - \frac{\hbar^2}{2M} \nabla_R^2 \psi(r_1, r_2, ... r_Z, R)$$

And the potential energy of the system looks like this:

$$PE = -\sum_{i=1}^{Z} \frac{Ze^2}{|r_i - R|} \psi(r_1, r_2, ... r_Z, R) + \sum_{i>j} \frac{e^2}{|r_i - r_j|} \psi(r_1, r_2, ... r_Z, R)$$

So adding the two preceding equations, here's what you get for the total energy (E = KE + PE) of a multi-particle atom:

$$E\psi(r_1, r_2, ... r_Z, R) = \sum_{i=1}^{Z} \frac{-\hbar^2}{2m_i} \nabla_i^2 \psi(r_1, r_2, ... r_Z, R) - \frac{\hbar^2}{2M} \nabla_R^2 \psi(r_1, r_2, ... r_Z, R)$$
$$- \sum_{i=1}^{Z} \frac{Ze^2}{|r_i - R|} \psi(r_1, r_2, ... r_Z, R) + \sum_{i>j} \frac{e^2}{|r_i - r_j|} \psi(r_1, r_2, ... r_Z, R)$$

Okay, now that looks like a proper mess. Want to win the Nobel prize in physics? Just come up with the general solution to the preceding equation. As is always the case when you have a multi-particle system in which the particles interact with each other, you can't split this equation into a system of N independent equations.

In cases where the N particles of a multi-particle system *don't* interact with each other, where you can disconnect the Schrödinger equation into a set of N independent equations, solutions may be possible. But when the particles interact and the Schrödinger equation depends on those interactions, you can't solve that equation for any significant number of particles.

However, that doesn't mean all is lost by any means. You can still say plenty about equations like this one if you're clever — and it all starts with an examination of the symmetry of the situation, which I discuss next.

# A Super-Powerful Tool: Interchange Symmetry

Even though finding general solutions for equations like the one for the total energy of a multi-particle atom (in the preceding section) is impossible, you can still see what happens when you exchange particles with each other — and the results are very revealing. This section covers the idea of interchange symmetry.

## Order matters: Swapping particles with the exchange operator

You can determine what happens to the wave function when you swap two particles. Whether the wave function is symmetric under such operations gives you insight into whether two particles can occupy the same quantum state. This section discusses swapping particles and looking at symmetric and antisymmetric functions.

Take a look at the general wave function for N particles:

$$\psi(r_1, r_2, ..., r_i, ..., r_j, ..., r_N)$$

***Note:*** In this chapter, I talk about symmetry in terms of the location coordinate, $r$, to keep things simple, but you can also consider other quantities, such as spin, velocity, and so on. That wouldn't make this discussion any different, because you can wrap all of a particle's quantum measurements — location, velocity, speed, and so on — into a single quantum state, which you can call $\xi$. Doing so would make the general wave function for N particles into this: $\psi(\xi_1, \xi_2, ..., \xi_i, ..., \xi_j, ..., \xi_N)$. But as I said, this section just considers the wave function $\psi(r_1, r_2, ..., r_i, ..., r_j, ..., r_N)$ to keep things simple.

Now imagine that you have an exchange operator, $P_{ij}$, that exchanges particles $i$ and $j$. In other words,

$$P_{ij}\psi(r_1, r_2, ..., r_i, ..., r_j, ..., r_N) = \psi(r_1, r_2, ..., r_j, ..., r_i, ..., r_N)$$

And $P_{ij} = P_{ji}$, so

$$P_{ij}\psi(r_1, r_2, ..., r_i, ..., r_j, ..., r_N) = \psi(r_1, r_2, ..., r_j, ..., r_i, ..., r_N)$$
$$= P_{ji}\psi(r_1, r_2, ..., r_i, ..., r_j, ..., r_N)$$

Also, note that applying the exchange operator twice just puts the two exchanged particles back where they were originally, so $P_{ij}^2 = 1$. Here's what that looks like:

$$P_{ij}P_{ij}\psi(r_1, r_2, ..., r_i, ..., r_j, ..., r_N) = P_{ij}\psi(r_1, r_2, ..., r_j, ..., r_i, ..., r_N)$$
$$= \psi(r_1, r_2, ..., r_i, ..., r_j, ..., r_N)$$

However, in general, $P_{ij}$ and $P_{lm}$ (where $ij \neq lm$) do not commute. That is, $P_{ij}P_{lm} \neq P_{lm}P_{ij}$ ($ij \neq lm$). Therefore, $[P_{ij}, P_{lm}] \neq 0$ ($ij \neq lm$). For example, say you have four particles whose wave function is

$$\psi(r_1, r_2, r_3, r_4) = \frac{r_1 r_3}{r_4} e^{r_2}$$

Apply the exchange operators $P_{12}$ and $P_{14}$ to see whether $P_{12}P_{14}$ equals $P_{14}P_{12}$. Here's $P_{14}\psi(r_1, r_2, r_3, r_4)$:

$$P_{14}\psi(r_1, r_2, r_3, r_4) = \frac{r_4 r_3}{r_1} e^{r_2}$$

And here's what $P_{12}P_{14}\psi(r_1, r_2, r_3, r_4)$ looks like:

$$P_{12}P_{14}\psi(r_1, r_2, r_3, r_4) = \frac{r_2 r_3}{r_1} e^{r_4}$$

Okay. Now take a look at $P_{14}P_{12}\psi(r_1, r_2, r_3, r_4)$. Here's $P_{12}\psi(r_1, r_2, r_3, r_4)$:

$$P_{12}\psi(r_1, r_2, r_3, r_4) = \frac{r_2 r_3}{r_4} e^{r_1}$$

And here's what $P_{14}P_{12}\psi(r_1, r_2, r_3, r_4)$ looks like:

$$P_{14}P_{12}\psi(r_1, r_2, r_3, r_4) = \frac{r_4 r_3}{r_2} e^{r_1}$$

As you can see by comparing $P_{12}P_{14}\psi(r_1,r_2,r_3,r_4) = \frac{r_2 r_3}{r_1}e^{r_4}$ and this last equation,

$P_{12}P_{14}\,\psi(r_1, r_2, r_3, r_4) \neq P_{14}P_{12}\,\psi(r_1, r_2, r_3, r_4)$. In other words, the order in which you apply exchange operators *matters*.

# Classifying symmetric and antisymmetric wave functions

$P_{ij}^2 = 1$ (see the preceding section), so note that if a wave function is an eigenfunction of $P_{ij}$, then the possible eigenvectors are 1 and –1. That is, for $\psi(r_1, r_2, ..., r_i, ..., r_j, ..., r_N)$ an eigenfunction of $P_{ij}$ looks like

$$P_{ij}\psi(r_1, r_2, ..., r_i, ..., r_j, ..., r_N) = \psi(r_1, r_2, ...r_i, ..., r_j, ..., r_N)$$
$$\text{or} -\psi(r_1, r_2, ..., r_i, ..., r_j, ..., r_N)$$

That means there are two kinds of eigenfunctions of the exchange operator:

- **Symmetric eigenfunctions:** $P_{ij}\psi_s(r_1, r_2, ..., r_i, ..., r_j, ..., r_N) = \psi_s(r_1, r_2, ..., r_i, ..., r_j, ..., r_N)$

- **Antisymmetric eigenfunctions:** $P_{ij}\psi_a(r_1, r_2, ..., r_i, ..., r_j, ..., r_N) = -\psi_a(r_1, r_2, ..., r_i, ..., r_j, ..., r_N)$

Now take a look at some symmetric and some antisymmetric eigenfunctions. How about this one — is it symmetric or antisymmetric?

$$\psi_1(r_1, r_2) = (r_1 - r_2)^2$$

You can apply the exchange operator $P_{12}$:

$$P_{12}\,\psi_1(r_1, r_2) = (r_2 - r_1)^2$$

Note that because $(r_1 - r_2)^2 = (r_2 - r_1)^2$, $\psi_1(r_1, r_2)$ is a symmetric wave function; that's because $P_{12}\,\psi_1(r_1, r_2) = \psi_1(r_1, r_2)$.

How about this wave function?

$$\psi_2(r_1, r_2) = \frac{r_1^2 + r_2^2}{(r_1 - r_2)^2}$$

Again, apply the exchange operator, $P_{12}$:

$$P_{12}\psi_2(r_1,r_2) = \frac{r_2^2 + r_1^2}{(r_1 - r_2)^2}$$

Okay, but because $\frac{r_2^2 + r_1^2}{(r_2 - r_1)^2} = \frac{r_1^2 + r_2^2}{(r_1 - r_2)^2}$ , you know that $P_{12}\psi_2(r_1, r_2) = \psi_2(r_1, r_2)$,

so $\psi_2(r_1, r_2)$ is symmetric.

Here's another one:

$$\psi_3(r_1 - r_2) = \frac{5(r_1 - r_2)}{(r_1 - r_2)^2}$$

Now apply $P_{12}$:

$$P_{12}\psi_3(r_1,r_2) = \frac{5(r_2 - r_1)}{(r_2 - r_1)^2}$$

How does that equation compare to the original one? Well, $\frac{5(r_2 - r_1)}{(r_2 - r_1)^2} = \frac{-5(r_1 - r_2)}{(r - r)^2}$ , so $P_{12}\psi_3(r_1, r_2) = -\psi_3(r_1, r_2)$. Therefore, $\psi_3(r_1, r_2)$ is antisymmetric.

What about this one?

$$\psi_4(r_1,r_2) = \frac{r_1 r_2}{(r_1 - r_2)^2} + r_1^2 + r_2^2$$

To find out, apply $P_{12}$:

$$P_{12}\psi_4(r_1,r_2) = \frac{r_2 r_1}{(r_2 - r_1)^2} + r_2^2 + r_1^2$$

All right — how's this compare with the original equation?

$$\frac{r_1 r_2}{(r_1 - r_2)^2} + r_1^2 + r_2^2 = \frac{r_2 r_1}{(r_2 - r_1)^2} + r_2^2 + r_1^2$$

Okay — $\psi_4(r_1, r_2)$ is symmetric.

You may think have this process down pretty well, but what about this next wave function?

$$\psi_5\left(r_1, r_2\right) = \frac{r_1 r_2}{\left(r_1 - r_2\right)^2} + r_1^{\ 2} - r_2^{\ 2}$$

Start by applying $P_{12}$:

$$P_{12}\psi_5\left(r_1, r_2\right) = \frac{r_2 + r_1}{\left(r_2 - r_1\right)^2} + r_2^{\ 2} - r_1^{\ 2}$$

So how do these two equations compare?

$$\frac{r_1 r_2}{\left(r_1 - r_2\right)^2} + r_1^{\ 2} - r_2^{\ 2} \neq \frac{r_2 r_1}{\left(r_2 - r_1\right)^2} + r_2^{\ 2} - r_1^{\ 2}$$

That is, $\psi_5(r_1, r_2)$ is neither symmetric nor antisymmetric. In other words, $\psi_5(r_1, r_2)$ is not an eigenfunction of the $P_{12}$ exchange operator.

# Floating Cars: Tackling Systems of Many Distinguishable Particles

All right, if you've been reading this chapter from the start, you pretty much have the idea of swapping particles down. Now you look at systems of particles that you can distinguish — that is, systems of identifiably different particles. As you see in this section, you can decouple such systems into linearly independent equations.

Suppose you have a system of many different types of cars floating around in space. You can distinguish all those cars because they're all different — they have different masses, for one thing.

Now say that each car interacts with its own potential — that is, the potential that any one car sees doesn't depend on any other car. That means that the potential for all cars is just the sum of the individual potentials each car sees, which looks like this, assuming you have N cars:

$$PE = V\left(r_1, r_2, \ldots, r_N\right) = \sum_{i=1}^{N} V\left(r_i\right)$$

Being able to cut the potential energy up into a sum of independent terms like this makes life a lot easier. Here's what the Hamiltonian looks like:

$$H\psi\left(r_1,r_2,...,r_N\right)=\sum_{i=1}^{N}\left[\frac{-\hbar^2}{2m_i}\nabla_i^2+V_i\left(r_i\right)\right]\psi\left(r_1,r_2,...,r_N\right)$$

Notice how much simpler this equation is than the Hamiltonian for the hydrogen atom which I give you here:

$$H\psi\left(r_1,r_2,...,r_z,\boldsymbol{R}\right)=\sum_{i=1}^{z}\frac{-\hbar^2}{2m_i}\nabla_i^2\psi\left(r_1,r_2,...,r_z,\boldsymbol{R}\right)-\frac{\hbar^2}{2M}\nabla_R^2\psi\left(r_1,r_2,...,r_z,\boldsymbol{R}\right)$$

$$-\sum_{i=1}^{z}\frac{Ze^2}{\left|r_i-R\right|}\psi\left(r_1,r_2,...,r_z,\boldsymbol{R}\right)+\sum_{i>j}\frac{e^2}{\left|r_i-r_j\right|}\psi\left(r_1,r_2,...,r_z,\boldsymbol{R}\right)$$

Note that you can separate the previous equation for the potential of all cars into N different equations:

$$\frac{-\hbar^2}{2m_i}\nabla_i^2\psi_i\left(r_i\right)+V_i\left(r_i\right)\psi_i\left(r_i\right)=E_i\psi_i\left(r_i\right)$$

And the total energy is just the sum of the energies of the individual cars:

$$E=\sum_{i=1}^{N}E_i$$

And the wave function is just the product of the individual wave functions:

$$\psi_{n_1,n_2,...,n_N}\left(r_1,r_2,...,r_N\right)=\prod_{i=1}^{N}\psi_{n_i}\left(r_i\right)$$

where the $\Pi$ symbol is just like $\Sigma$, except it stands for a product of terms, not a sum, and $n_i$ refers to all the quantum numbers of the $i$th particle.

As you can see, when the particles you're working with are distinguishable and subject to independent potentials, the problem of handling many of them becomes simpler. You can break the system up into N independent one-particle systems. The total energy is just the sum of the individual energies of each particle. The Schrödinger equation breaks down into N different equations. And the wave function ends up just being the product of the wave functions of the N different particles.

Take a look at an example. Say you have four particles, each with a different mass, in a square well. You want to find the energy and the wave function of this system. Here's what the potential of the square well looks like this for each of the four noninteracting particles:

$$V_i(x_i) = \begin{vmatrix} 0 \text{ for } 0 \le x_i \le a \\ \infty \text{ for } x_i > a \\ \infty \text{ for } x_i < 0 \end{vmatrix}$$

Here's what the Schrödinger equation looks like:

$$E\psi_{n_1,n_2,n_3,n_4}(x_1,x_2,x_3,x_4) = \sum_{i=1}^{4} \frac{-\hbar^2}{2m_i} \frac{d^2}{dx_i^2} \psi_{n_1,n_2,n_3,n_4}(x_1,x_2,x_3,x_4)$$

You can separate the preceding equation into four one-particle equations:

✔ $E_1\psi_{n_1}(x_1) = \frac{-\hbar^2}{2m_1} \frac{d^2}{dx_1^2} \psi_{n_1}(x_1)$

✔ $E_2\psi_{n_2}(x_2) = \frac{-\hbar^2}{2m_2} \frac{d^2}{dx_2^2} \psi_{n_2}(x_2)$

✔ $E_3\psi_{n_3}(x_3) = \frac{-\hbar^2}{2m_3} \frac{d^2}{dx_3^2} \psi_{n_3}(x_3)$

✔ $E_4\psi_{n_4}(x_4) = \frac{-\hbar^2}{2m_4} \frac{d^2}{dx_4^2} \psi_{n_4}(x_4)$

I've already solved such one-dimensional problems in Chapter 3. The energy levels are

$$E_i = \frac{\hbar^2 \pi^2 n_i^2}{2m_i a^2}$$

And because the total energy is the sum of the individual energies is $E = \sum_{i=1}^{4} E_i$, the energy in general is

$$E = \frac{\hbar^2 \pi^2}{2a^2} \left[ \frac{n_1^2}{m_1} + \frac{n_2^2}{m_2} + \frac{n_3^2}{m_3} + \frac{n_4^2}{m_4} \right]$$

So here's the energy of the ground state — where all particles are in their ground states, $n_1 = n_2 = n_3 = n_4 = 1$:

$$E = \frac{\hbar^2 \pi^2}{2a^2} \left[ \frac{1}{m_1} + \frac{1}{m_2} + \frac{1}{m_3} + \frac{1}{m_4} \right]$$

For a one-dimensional system with a particle in a square well, the wave function is

$$\psi_i(x) = \frac{2^{1/2}}{a^{1/2}} \sin\left(\frac{n_i \pi}{a} x_i\right)$$

The wave function for the four-particle system is just the product of the individual wave functions, so it looks like this:

$$\psi_{n_1,n_2,n_3,n_4}(x_1,x_2,x_3,x_4) = \frac{4}{a^2} \sin\left(\frac{n_1 \pi}{a} x_1\right) \sin\left(\frac{n_2 \pi}{a} x_2\right) \sin\left(\frac{n_3 \pi}{a} x_3\right) \sin\left(\frac{n_4 \pi}{a} x_4\right)$$

For example, for the ground state, $n_1 = n_2 = n_3 = n_4 = 1$, you have

$$\psi_{1,1,1,1}(x_1,x_2,x_3,x_4) = \frac{4}{a^2} \sin\left(\frac{\pi}{a} x_1\right) \sin\left(\frac{\pi}{a} x_2\right) \sin\left(\frac{\pi}{a} x_3\right) \sin\left(\frac{\pi}{a} x_4\right)$$

So as you can see, systems of N independent, distinguishable particles are often susceptible to solution — all you have to do is to break them up into N independent equations.

# Juggling Many Identical Particles

When the particles in a multi-particle system are all indistinguishable, that's when the real adventure begins. When you can't tell the particles apart, how can you tell which one's where? This section explains what happens.

## Losing identity

Say you have a bunch of pool balls and you want to look at them classically. You can paint each pool ball differently, and then, even as they hurtle around the pool table, you're able to distinguish them — seven ball in the corner pocket, and that sort of thing. Classically, identical particles retain their individuality. You can still tell them apart.

The same isn't true quantum mechanically, because you can't locate particles with absolute precision. So if you were to have a bunch of electrons, you'd quickly lose track of which one was which — you can't paint them, as you can pool balls.

For example, look at the scenario in Figure 10-3. There, two electrons are colliding and bouncing apart. Seems like keeping track of the two electrons would be easy.

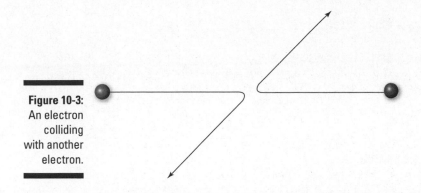

**Figure 10-3:**
An electron
colliding
with another
electron.

But now look at the scenario in Figure 10-4 — the electrons could've bounced like that, not like the bounce shown in Figure 10-3. And you'd never know it.

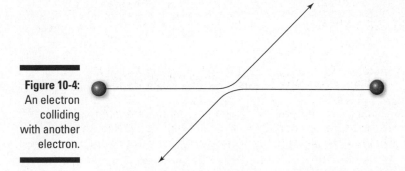

**Figure 10-4:**
An electron
colliding
with another
electron.

So which electron is which? From the experimenter's point of view, you can't tell. You can place detectors to catch the electrons, but you can't determine which of the incoming electrons ended up in which detector, because of the two possible scenarios in Figures 10-3 and 10-4.

Quantum mechanically, identical particles don't retain their individuality in terms of any measurable, observable quantity. You lose the individuality of identical particles as soon as you mix them with similar particles. This idea holds true for any N-particle system. As soon as you let N identical particles interact, you can't say which exact one is at $r_1$ or $r_2$ or $r_3$ or $r_4$ and so on.

# Symmetry and antisymmetry

In practical terms, the loss of individuality among identical particles means that the probability density remains unchanged when you exchange particles. For example, if you were to exchange electron 10,281 with electron 59,830, you'd still have the same probability that an electron would occupy $d^3r_{10,281}$ and $d^3r_{59,830}$.

Here's what this idea looks like mathematically ($r$ and $s$ are the location and spins of the particles):

$$|\psi(r_1s_1, r_2s_2, ..., r_is_i, ..., r_js_j, ..., r_Ns_N)|^2 = |\psi(r_1s_1, r_2s_2, ..., r_js_j, ..., r_is_i, ..., r_Ns_N)|^2$$

The preceding equation means that

$$\psi(r_1s_1, r_2s_2, ..., r_is_i, ..., r_js_j, ..., r_Ns_N) = \pm\psi(r_1s_1, r_2s_2, ..., r_js_j, ..., r_is_i, ..., r_Ns_N)$$

So the wave function of a system of N identical particles must be either symmetric or antisymmetric when you exchange two particles. Spin turns out to be the deciding factor:

- **Antisymmetric wave function:** If the particles have half-odd-integral spin ($1/2$, $3/2$, and so on), then this is how the wave function looks under exchange of particles:

  $$\psi(r_1s_1, r_2s_2, ..., r_is_i, ..., r_js_j, ..., r_Ns_N) = -\psi(r_1s_1, r_2s_2, ..., r_js_j, ..., r_is_i, ..., r_Ns_N)$$

- **Symmetric wave function:** If the particles have integral spin (0, 1, and so on), this is how the wave function looks under exchange of particles:

  $$\psi(r_1s_1, r_2s_2, ..., r_is_i, ..., r_js_j, ..., r_Ns_N) = \psi(r_1s_1, r_2s_2, ..., r_js_j, ..., r_is_i, ..., r_Ns_N)$$

Having symmetric or antisymmetric wave functions leads to some different physical behavior, depending on whether the wave function is symmetric or antisymmetric.

In particular, particles with integral spin, such as photons or pi mesons, are called *bosons*. And particles with half-odd-integral spin, such as electrons, protons, and neutrons, are called *fermions*. The behavior of systems of fermions is very different from the behavior of systems of bosons.

# Exchange degeneracy: The steady Hamiltonian

The Hamiltonian, which you can represent like this

$$H(r_1s_1, r_2s_2, ..., r_is_i, ..., r_js_j, ..., r_Ns_N)$$

doesn't vary under exchange of two identical particles. In other words, the Hamiltonian is invariant here, no matter how many identical particles you exchange. That's called *exchange degeneracy*, and mathematically, it looks like this:

$$H(r_1s_1, r_2s_2, ..., r_is_i, ..., r_js_j, ..., r_Ns_N) = H(r_1s_1, r_2s_2, ..., r_js_j, ..., r_is_i, ..., r_Ns_N)$$

That means, incidentally, that the exchange operator, $P_{ij}$, is an invariant of the motion because it commutes with the Hamiltonian:

$$[H, P_{ij}] = 0$$

## Name that composite: Grooving with the symmetrization postulate

In the earlier section titled "Symmetry and antisymmetry," I show that the wave function of a system of N particles is either symmetric or antisymmetric under the exchange of two particles:

- **Symmetric:** $\psi(r_1s_1, r_2s_2, ..., r_is_i, ..., r_js_j, ..., r_Ns_N) = \psi(r_1s_1, r_2s_2, ..., r_js_j, ..., r_is_i, ..., r_Ns_N)$
- **Antisymmetric:** $\psi(r_1s_1, r_2s_2, ..., r_is_i, ..., r_js_j, ..., r_Ns_N) = -\psi(r_1s_1, r_2s_2, ..., r_js_j, ..., r_is_i, ..., r_Ns_N)$

This turns out to be the basis of the *symmetrization postulate,* which says that in systems of N identical particles, only states that are symmetric or antisymmetric exist — and it says that states of mixed symmetry don't exist.

The symmetrization postulate also says, as observed from nature, that

- Particles with half-odd-integral states ($^1/_2$, $^3/_2$, $^5/_2$, ...) are *fermions,* and they have antisymmetric states under the interchange of two particles.

- Particles with integral spin (0, 1, 2, ...) are *bosons,* and they have symmetric states under the interchange of two particles.

So the wave function of N fermions is completely antisymmetric, and the wave function of N bosons is completely symmetric.

Determining whether a particle is a fermion or a boson may seem like an easy task — just look it up. Electrons, protons, and neutrons are fermions, for example, with half-odd-integral spin. And photons, pi mesons, and so on are bosons, with integral spins.

But what if the particle you're studying is a composite particle? What if, for example, you have an *alpha particle,* which is made up of two protons and two neutrons? Is that a fermion or a boson?

In fact, protons and neutrons themselves are made up of three quarks, and pi mesons are made up of two — and quarks have spin $1/2$.

Composites can be either fermions or bosons — it all depends on whether the spin of the composite particle ends up being half-odd-integral or integral. If the composite particle's spin is $1/2$, $3/2$, $5/2$, and so on, then the composite particle is a fermion. If the composite particle's spin is 0, 1, 2, and so on, then the composite particle is a boson.

In general, if the composite particle is made up of an odd number of fermions, then it's a fermion. Otherwise, it's a boson. So for example, because quarks are fermions and because nucleons such as protons and neutrons are made up of three quarks, those nucleons end up being fermions. But because pi mesons are made up of two quarks, they end up being bosons. The alpha particle, which consists of two protons and two neutrons, is a boson. You can even consider whole atoms to be composite particles. For example, consider the hydrogen atom: That atom is made up of a proton (a fermion) and an electron (another fermion), so that's two fermions. And that makes the hydrogen atom a boson.

# Building Symmetric and Antisymmetric Wave Functions

Many of the wave functions that are solutions to physical setups like the square well aren't inherently symmetric or antisymmetric; they're simply *asymmetric.* In other words, they have no definite symmetry. So how do you end up with symmetric or antisymmetric wave functions?

The answer is that you have to create them yourself, and you do that by adding together asymmetric wave functions. For example, say that you have an asymmetric wave function of two particles, $\psi(r_1 s_1, r_2 s_2)$.

To create a symmetric wave function, add together $\psi(r_1 s_1, r_2 s_2)$ and the version where the two particles are swapped, $\psi(r_2 s_2, r_1 s_1)$. Assuming that $\psi(r_1 s_1, r_2 s_2)$ and $\psi(r_2 s_2, r_1 s_1)$ are normalized, you can create a symmetric wave function using these two wave functions this way — just by adding the wave functions:

$$\psi_s\left(r_1 s_1, r_2 s_2\right) = \frac{1}{\sqrt{2}}\left[\psi\left(r_1 s_1, r_2 s_2\right) + \psi\left(r_2 s_2, r_1 s_1\right)\right]$$

You can make an antisymmetric wave function by subtracting the two wave functions:

$$\psi_a\left(r_1 s_1, r_2 s_2\right)=\frac{1}{\sqrt{2}}\left[\psi\left(r_1 s_1, r_2 s_2\right)-\left(r_2 s_2, r_1 s_1\right)\right]$$

This process gets rapidly more complex the more particles you add, however, because you have to interchange all the particles. For example, what would a symmetric wave function based on the asymmetric three-particle wave function $\psi(r_1 s_1, r_2 s_2, r_3 s_3)$ look like? Why, it'd look like this:

$$\psi_s\left(r_1 s_1, r_2 s_2, r_3 s_3\right)=\frac{1}{\sqrt{6}}\Big[\psi\left(r_1 s_1, r_2 s_2, r_3 s_3\right)+\psi\left(r_1 s_1, r_3 s_3, r_2 s_2\right)$$
$$+\psi\left(r_2 s_2, r_3 s_3, r_1 s_1\right)+\psi\left(r_2 s_2, r_1 s_1, r_3 s_3\right)$$
$$+\psi\left(r_3 s_3, r_1 s_1, r_2 s_2\right)+\psi\left(r_3 s_3, r_2 s_2, r_1 s_1\right)\Big]$$

And how about the antisymmetric wave function? That looks like this:

$$\psi_s\left(r_1 s_1, r_2 s_2, r_3 s_3\right)=$$
$$\frac{1}{\sqrt{6}}\begin{bmatrix}\psi\left(r_1 s_1, r_2 s_2, r_3 s_3\right)-\psi\left(r_1 s_1, r_3 s_3, r_2 s_2\right)+\psi\left(r_2 s_2, r_3 s_3, r_1 s_1\right)\\ -\psi\left(r_2 s_2, r_1 s_1, r_3 s_3\right)+\psi\left(r_3 s_3, r_1 s_1, r_2 s_2\right)-\psi\left(r_3 s_3, r_2 s_2, r_1 s_1\right)\end{bmatrix}$$

And in this way, at least theoretically, you can create symmetric and antisymmetric wave functions for any system of N particles.

# Working with Identical Noninteracting Particles

Working with identical noninteracting particles makes life easier because you can treat the equations individually instead of combining them into one big mess. Say you have a system of N identical particles, each of which experiences the same potential. You can separate the Schrödinger equation into N identical single-particle equations:

$$\frac{-\hbar^2}{2m_i}\nabla_i^2\psi_i\left(r_i\right)+V_i\left(r_i\right)\psi_i\left(r_i\right)=E_i\psi_i\left(r_i\right)$$

And the total energy is just the sum of the energies of the individual particles:

$$E = \sum_{i=1}^{N} E_i$$

But now look at the wave function for the system. Earlier in the chapter (see "Floating Cars: Tackling Systems of Many Distinguishable Particles"), you consider the wave function of a system of N distinguishable particles and come up with the product of all the individual wave functions:

$$\psi_{n_1, n_2, \ldots, n_N}\left(r_1, r_2, \ldots, r_N\right) = \prod_{i=1}^{N} \psi_{n_i}\left(r_i\right)$$

However, that equation doesn't work with identical particles because you can't say that particle 1 is in state $\psi_1(r_1)$, particle 2 is in state $\psi_2(r_2)$, and so on — they're identical particles here, not distinguishable particles as before.

The other reason this equation doesn't work here is that it has no inherent symmetry — and systems of N identical particles must have a definite symmetry. So instead of simply multiplying the wave functions, you have to be a little more careful.

## Wave functions of two-particle systems

How do you create symmetric and antisymmetric wave functions for a two-particle system? Start with the single-particle wave functions (see the earlier section "Building Symmetric and Antisymmetric Wave Functions"):

$$\psi_s\left(r_1 s_1, r_2 s_2\right) = \frac{1}{\sqrt{2}}\left[\psi\left(r_1 s_1, r_2 s_2\right) + \psi\left(r_2 s_2, r_1 s_1\right)\right]$$

$$\psi_a\left(r_1 s_1, r_2 s_2\right) = \frac{1}{\sqrt{2}}\left[\psi\left(r_1 s_1, r_2 s_2\right) - \left(r_2 s_2, r_1 s_1\right)\right]$$

By analogy, here's the symmetric wave function, this time made up of two single-particle wave functions:

$$\psi_s\left(r_1 s_1, r_2 s_2\right) = \frac{1}{\sqrt{2}}\left[\psi_{n_1}\left(r_1 s_1\right)\psi_{n_2}\left(r_2 s_2\right) + \psi_{n_1}\left(r_2 s_2\right)\psi_{n_2}\left(r_1 s_1\right)\right]$$

And here's the antisymmetric wave function, made up of the two single-particle wave functions:

$$\psi_a\left(r_1 s_1, r_2 s_2\right) = \frac{1}{\sqrt{2}}\left[\psi_{n_1}\left(r_1 s_1\right)\psi_{n_2}\left(r_2 s_2\right) - \psi_{n_1}\left(r_2 s_2\right)\psi_{n_2}\left(r_1 s_1\right)\right]$$

where $n_i$ stands for all the quantum numbers of the $i$th particle.

Note in particular that $\psi_a(r_1 s_1, r_2 s_2) = 0$ when $n_1 = n_2$; in other words, the antisymmetric wave function vanishes when the two particles have the same set of quantum numbers — that is, when they're in the same quantum state. That idea has important physical ramifications.

You can also write $\psi_s(r_1 s_1, r_2 s_2)$ like this, where P is the permutation operator, which takes the permutation of its argument:

$$\psi_s\left(r_1 s_1, r_2 s_2\right) = \frac{1}{\sqrt{2!}}\sum_P P\psi_{n_1}\left(r_1 s_1\right)\psi_{n_2}\left(r_2 s_2\right)$$

And also note that you can write $\psi_a(r_1 s_1, r_2 s_2)$ like this:

$$\psi_a\left(r_1 s_1, r_2 s_2\right) = \frac{1}{\sqrt{2!}}\sum_P (-1)^P P\psi_{n_1}\left(r_1 s_1\right)\psi_{n_2}\left(r_2 s_2\right)$$

where the term $(-1)^P$ is 1 for even permutations (where you exchange both $r_1 s_1$ and $r_2 s_2$ and also $n_1$ and $n_2$) and –1 for odd permutations (where you exchange $r_1 s_1$ and $r_2 s_2$ but not $n_1$ and $n_2$; or you exchange $n_1$ and $n_2$ but not $r_1 s_1$ and $r_2 s_2$).

In fact, people sometimes write $\psi_a(r_1 s_1, r_2 s_2)$ in determinant form like this:

$$\psi_a\left(r_1 s_1, r_2 s_2\right) = \frac{1}{\sqrt{2!}}\det\begin{vmatrix} \psi_{n_1}\left(r_1 s_1\right) & \psi_{n_1}\left(r_2 s_2\right) \\ \psi_{n_2}\left(r_1 s_1\right) & \psi_{n_2}\left(r_2 s_2\right) \end{vmatrix}$$

Note that this determinant is zero if $n_1 = n_2$.

# Wave functions of three-or-more-particle systems

Now you get to put together the wave function of a system of three particles from single-particle wave functions.

The symmetric wave function looks like this:

$$\psi_s\left(r_1 s_1, r_2 s_2, r_3 s_3\right) = \frac{1}{\sqrt{3!}} \sum_P P \psi_{n_1}\left(r_1 s_1\right) \psi_{n_2}\left(r_2 s_2\right) \psi_{n_3}\left(r_3 s_3\right)$$

And the antisymmetric wave function looks like this:

$$\psi_a\left(r_1 s_1, r_2 s_2, r_2 s_2\right) = \frac{1}{\sqrt{3!}} \sum_P (-1)^P P \psi_{n_1} 1\left(r_1 s_1\right) \psi_{n_2}\left(r_2 s_2\right) \psi_{n_3}\left(r_3 s_3\right)$$

This asymmetric wave function goes to zero if any two single particles have the same set of quantum numbers ($n_i = n_j, i \neq j$).

How about generalizing this to systems of N particles? If you have a system of N particles, the symmetric wave function looks like this:

$$\psi_s\left(r_1 s_1, r_2 s_2, \ldots, r_N s_N\right) = \frac{1}{\sqrt{N!}} \sum_P P \psi_{n_1}\left(r_1 s_1\right) \psi_{n_2}\left(r_2 s_2\right) \ldots \psi_{n_N}\left(r_N s_N\right)$$

And the antisymmetric wave function looks like this:

$$\psi_a\left(r_1 s_1, r_2 s_2, \ldots, r_N s_N\right) = \frac{1}{\sqrt{N!}} \sum_P (-1)^P P \psi_{n_1}\left(r_1 s_1\right) \psi_{n_2}\left(r_2 s_2\right) \ldots \psi_{n_N}\left(r_N s_N\right)$$

The big news is that the antisymmetric wave function for N particles goes to zero if any two particles have the same quantum numbers ($n_i = n_j, i \neq j$). And that has a big effect in physics, as you see next.

# It's Not Come One, Come All: The Pauli Exclusion Principle

The antisymmetric wave function vanishes if any two particles in an N-particle system have the same quantum numbers. Because fermions are the type of particles that have antisymmetric wave functions, that's the equivalent of saying that in a system of N particles, no two fermions can have the same quantum numbers — that is, occupy the same state.

That idea, which Austrian physicist Wolfgang Pauli first formulated in 1925, is called the *Pauli exclusion principle.* The topic of discussion at that time was the atom, and the Pauli exclusion principle applied to the electrons (a type of fermion), which are present in all atoms.

The *Pauli exclusion principle* states that no two electrons can occupy the same quantum state inside a single atom. And that result is important for the structure of atoms. Instead of just piling on willy-nilly, electrons have to fill quantum states that aren't already taken. The same isn't true for bosons — for example, if you have a heap of alpha particles (bosons), they can all be in the same quantum state. Not so for fermions.

There are various quantum numbers that electrons can take in an atom — $n$ (the energy), $l$ (the angular momentum), $m$ (the $z$ component of the angular momentum), and $m_s$ (the $z$ component of spin). And using that information, you can construct the electron structure of atoms.

# Figuring out the Periodic Table

One of the biggest successes of the Schrödinger equation, together with the Pauli exclusion principle (see the preceding section), is explaining the electron structure of atoms.

The electrons in an atom have a shell structure, and they fill that structure based on the Pauli exclusion principle, which maintains that no two electrons can have the same state:

- ✔ The major shells are specified by the principal quantum number, $n$, corresponding to the distance of the electron from the nucleus.

- ✔ Shells, in turn, have subshells based on the orbital angular momentum quantum number, $l$.

- ✔ In turn, each subshell has subshells — called *orbitals* — which are based on the $z$ component of the angular momentum, $m$.

So each shell $n$ has n – 1 subshells, corresponding to l = 0, 1, 2, ..., $n$ – 1. And in turn, each subshell has $2l$ + 1 orbitals, corresponding to $m$ = –1, –$l$ + 1, ..., $l$ – 1, $l$.

Much as with the hydrogen atom, the various subshells ($l$ = 0, 1, 2, 3, 4, and so on) are called the *s, p, d, f, g, h,* and so on states. So, for example, for a given $n$, an $s$ state has one orbital ($m$ = 0), a $p$ state has three orbitals ($m$ = –1, 0, and 1), a $d$ state has five orbitals ($m$ = –2, –1, 0, 1, and 2), and so on.

In addition, due to the $z$ component of the spin, $m_s$ each orbital can contain two electrons — one with spin up, and one with spin down.

So how do the electrons, as fermions, fill the structure of an atom? Electrons can't fill a quantum state that's already been taken. For atoms in the ground state, electrons fill the orbitals in order of increasing energy. As soon as all of a subshell's orbitals are filled, the next electron goes on to the next subshell; and when the subshell is filled, the next electron goes on to the next shell, and so on.

Of course, as you fill the different electron shells, subshells, and orbitals, you end up with a different electron structure. And because interactions between electrons form the basis of chemistry, as electrons fill the successive quantum levels in various atoms, you end up with different chemical properties for those atoms — which set up the period (row) and group (column) organization of the periodic table.

# Part V
# Group Dynamics: Introducing Multiple Particles

The 5th Wave          By Rich Tennant

Einstein Working on One of His Concepts of Time – the "Good Time".

# In this part . . .

This part introduces you to working with multiple particles at the same time. Now, all the particles in the system can interact not only with an overall potential but also with each other. You see how to deal with atoms (electron and nucleus systems) here, as well as systems of many atoms. After all, the whole world is made up of many-particle systems. Good thing quantum physics is up to the task.

# Chapter 11

# Giving Systems a Push: Perturbation Theory

*P*roblems in quantum physics can become pretty tough pretty fast — another way of saying that, unfortunately, you just can't find exact solutions to many quantum physics problems. This is particularly the case when you merge two kinds of systems. For example, you may know all about how square wells work and all about how electrons in magnetic fields work, but what if you combine the two? The wave functions of each system, which you know exactly, are no longer applicable — you need some sort of mix instead.

Perturbation theory to the rescue! This theory lets you handle mixes of situations, as long as the interference isn't too strong. In this chapter, you explore time-independent perturbation theory and degenerate and nondegenerate Hamiltonians. You also look at some examples that place harmonic oscillators and hydrogen atoms in electric fields.

# Introducing Time-Independent Perturbation Theory

The idea behind time-independent perturbation theory is that you start with a known system — one whose wave functions you know and whose energy levels you know. Everything is all set up to this point. Then some new stimulus — a *perturbation* — comes along, disturbing the status quo. For example, you may apply an electrostatic or magnetic field to your known system, which changes that system somewhat.

Perturbation theory lets you handle situations like this — as long as the perturbation isn't too strong. In other words, if you apply a weak magnetic field to your known system, the energy levels will be mostly unchanged but with a correction. (**Note:** That's why it's called *perturbation theory* and not *drastic-interference theory.*) The change you make to the setup is slight enough so that you can calculate the resulting energy levels and wave functions as *corrections* to the fundamental energy levels and wave functions of the unperturbed system.

So what does it mean to talk of perturbations in physics terms? Say that you have this Hamiltonian:

$$H = H_0 + \lambda W \quad (\lambda << 1)$$

Here, $H_0$ is a known Hamiltonian, with known eigenfunctions and eigenvalues, and $\lambda W$ is the so-called perturbation Hamiltonian, where $\lambda << 1$ indicates that the perturbation Hamiltonian is small.

Finding the eigenstates of the Hamiltonian in this equation is what solving problems like this is all about — in other words, here's the problem you want to solve:

$$H|\psi_n\rangle = (H_0 + \lambda W)|\psi_n\rangle = E_n|\psi_n\rangle \quad (\lambda << 1)$$

The way you solve this equation depends on whether the exact, known solutions of $H_0$ are *degenerate* (that is, several states have the same energy) or nondegenerate. The next section solves the nondegenerate case.

# Working with Perturbations to Nondegenerate Hamiltonians

Start with the case in which the unperturbed Hamiltonian, $H_0$, has *nondegenerate* solutions. That is, for every state $|\phi_n>$, there's exactly one energy, $E_n$, that isn't the same as the energy for any other state: $H_0|\phi_n\rangle = E_n|\phi_n\rangle$ (just as a one-to-one function has only one $x$ value for any $y$). You refer to these nondegenerate energy levels of the unperturbed Hamiltonian as $E^{(0)}_n$ to distinguish them from the corrections that the perturbation introduces, so the equation becomes

$$H_0|\phi_n\rangle = E^{(0)}_n|\phi_n\rangle$$

From here on, I refer to the energy levels of the perturbed system as $E_n$.

The idea behind perturbation theory is that you can perform expansions based on the parameter $\lambda$ (which is much, much less than 1) to find the wave functions and energy levels of the perturbed system. In this section, you go up to terms in $\lambda^2$ in the expansions.

## A little expansion: Perturbing the equations

To find the energy of the perturbed system, $E_n$, start with the energy of the unperturbed system:

$$E_n = E^{(0)}{}_n + \ldots$$

Add the first-order correction to the energy, $\lambda E^{(1)}{}_n$:

$$E_n = E^{(0)}{}_n + \lambda E^{(1)}{}_n + \ldots \quad \left( \lambda \ll 1 \right)$$

And add the second-order correction to the energy, $\lambda^2 E^{(2)}{}_n$, as well:

$$E_n = E^{(0)}{}_n + \lambda E^{(1)}{}_n + \lambda^2 E^{(2)}{}_n + \ldots \quad \left( \lambda \ll 1 \right)$$

Now what about the wave function of the perturbed system, $|\psi_n\rangle$? Start with the wave function of the unperturbed system, $|\phi_n\rangle$:

$$\left| \psi_n \right\rangle = \left| \phi_n \right\rangle + \ldots$$

Add to it the first-order correction, $\lambda |\psi^{(1)}{}_n\rangle$:

$$\left| \psi_n \right\rangle = \left| \phi_n \right\rangle + \lambda \left| \psi^{(1)}{}_n \right\rangle + \ldots \quad \left( \lambda \ll 1 \right)$$

And then add to that the second-order correction to the wave function, $\lambda^2 |\psi^{(2)}{}_n\rangle$:

$$\left| \psi_n \right\rangle = \left| \phi_n \right\rangle + \lambda \left| \psi^{(1)}{}_n \right\rangle + \lambda^2 \left| \psi^{(2)}{}_n \right\rangle + \ldots \quad \left( \lambda \ll 1 \right)$$

Note that when $\lambda \to 0$, $E_n = E^{(0)}{}_n + \lambda E^{(1)}{}_n + \lambda^2 E^{(2)}{}_n + \ldots \quad \left( \lambda \ll 1 \right)$ becomes the unperturbed energy:

$$E_n = E^{(0)}{}_n$$

And $\left| \psi_n \right\rangle = \left| \phi_n \right\rangle + \lambda \left| \psi^{(1)}{}_n \right\rangle + \lambda^2 \left| \psi^{(2)}{}_n \right\rangle + \ldots \quad \left( \lambda \ll 1 \right)$ becomes the unperturbed wave function:

$$\left| \psi_n \right\rangle = \left| \phi_n \right\rangle$$

So your task is to calculate $E^{(1)}_n$ and $E^{(2)}_n$, as well as $\psi^{(1)}_n$ and $\psi^{(2)}_n$. So how do you do that in general? Time to start slinging some math. You start with three perturbed equations:

- **Hamiltonian:** $H|\psi_n\rangle = (H_0 + \lambda W)|\psi_n\rangle = E_n|\psi_n\rangle$  $(\lambda \ll 1)$
- **Energy levels:** $E_n = E^{(0)}_n + \lambda E^{(1)}_n + \lambda^2 E^{(2)}_n + \dots$  $(\lambda \ll 1)$
- **Wave functions:** $|\psi_n\rangle = |\phi_n\rangle + \lambda|\psi^{(1)}_n\rangle + \lambda^2|\psi^{(2)}_n\rangle + \dots$  $(\lambda \ll 1)$

Combine these three equations to get this jumbo equation:

$$(H_0 + \lambda W)\left(|\phi_n\rangle + \lambda|\psi^{(1)}_n\rangle + \lambda^2|\psi^{(2)}_n\rangle + \dots\right)$$
$$= \left(E^{(0)}_n + \lambda E^{(1)}_n + \lambda^2 E^{(2)}_n + \dots\right)\left(|\phi_n\rangle + \lambda|\psi^{(1)}_n\rangle + \lambda^2|\psi^{(2)}_n\rangle + \dots\right)  \quad (\lambda \ll 1)$$

# *Matching the coefficients of $\lambda$ and simplifying*

You can handle the jumbo equation in the preceding section by setting the coefficients of $\lambda$ on either side of the equal sign equal to each other.

Equating the zeroth order terms in $\lambda$ on either side of this equation, here's what you get:

$$H_0|\phi_n\rangle = E^{(0)}_n|\phi_n\rangle$$

Now for the first-order terms in $\lambda$; equating them on either side of the jumbo equation gives you

$$H_0|\psi^{(1)}_n\rangle + W|\phi_n\rangle = E^{(0)}_n|\psi^{(1)}_n\rangle + E^{(1)}_n|\phi_n\rangle$$

Now equate the coefficients of $\lambda^2$ in the jumbo equation, giving you

$$H_0|\psi^{(2)}_n\rangle + W|\psi^{(1)}_n\rangle = E^{(0)}_n|\psi^{(2)}_n\rangle + E^{(1)}_n|\psi^{(1)}_n\rangle + E^{(2)}_n|\phi_n\rangle$$

Okay, that's the equation you derive from the second order in $\lambda$. Now you have to solve for $E^{(1)}_n$, $E^{(2)}_n$, and so on using the zeroth-order, first-order, and second-order equations.

Start by noting that the unperturbed wave function, $|\phi_n>$ isn't going to be very different from the perturbed wave function, $|\psi_n>$, because the perturbation is small. That means that $\langle \phi_n | \psi_n \rangle \approx 1$. In fact, you can normalize $|\psi_n>$ so that $<\phi_n | \psi_n>$ is exactly equal to 1:

$$\langle \phi_n | \psi_n \rangle = 1$$

Given that $|\psi_n\rangle = |\phi_n\rangle + \lambda |\psi^{(1)}{}_n\rangle + \lambda^2 |\psi^{(2)}{}_n\rangle + \ldots$, the equation becomes

$$\lambda \langle \phi_n | \psi^{(1)}{}_n \rangle + \lambda^2 \langle \phi_n | \psi^{(2)}{}_n \rangle + \ldots = 0$$

And because the coefficients of $\lambda$ must both vanish, you get the following:

$$\langle \phi_n | \psi^{(1)}{}_n \rangle = \langle \phi_n | \psi^{(2)}{}_n \rangle = 0$$

This equation is useful for simplifying the math.

## Finding the first-order corrections

After matching the coefficients of $\lambda$ and simplifying (see the preceding section), you want to find the first-order corrections to the energy levels and the wave functions. Find the first-order correction to the energy, $E^{(1)}{}_n$, by multiplying $H_0 |\psi^{(1)}{}_n\rangle + W|\phi_n\rangle = E^{(0)}{}_n |\psi^{(1)}{}_n\rangle + E^{(1)}{}_n |\phi_n\rangle$ by $<\phi_n|$:

$$\langle \phi_n | H_0 | \psi^{(1)}{}_n \rangle + \langle \phi_n | W | \phi_n \rangle = \langle \phi_n | E^{(0)}{}_n | \psi^{(1)}{}_n \rangle + \langle \phi_n | E^{(1)}{}_n | \phi_n \rangle$$

You can use $H|\psi_n\rangle = (H_0 + \lambda W)|\psi_n\rangle = E_n |\psi_n\rangle$ $(\lambda \ll 1)$ to simplify this to

$$E^{(1)}{}_n = \langle \phi_n | W | \phi_n \rangle$$

Swell, that's the expression you use for the first-order correction, $E^{(1)}{}_n$.

Now look into finding the first-order correction to the wave function, $|\psi^{(1)}{}_n>$. You can multiply the wave-function equation by this next expression, which is equal to 1:

$$\sum_m |\phi_m\rangle\langle \phi_m|$$

So you have

$$\left| \psi^{(1)}_{\,n} \right\rangle = \left( \sum_m \left| \phi_m \right\rangle \left\langle \phi_m \right| \right) \left| \psi^{(1)}_{\,n} \right\rangle$$

$$= \sum_m \left\langle \phi_m \left| \psi^{(1)}_{\,n} \right\rangle \right| \phi_m \right\rangle \qquad m \neq n$$

Note that the $m = n$ term is zero because $\langle \phi_n | \psi^{(1)}_{\,n} \rangle = 0$.

So what is $\langle \phi_m | \psi^{(1)}_{\,n} \rangle$? You can find out by multiplying the first-order correction, $H_0 \left| \psi^{(1)}_{\,n} \right\rangle + W \left| \phi_n \right\rangle = E^{(0)}_{\,n} \left| \psi^{(1)}_{\,n} \right\rangle + E^{(1)}_{\,n} \left| \phi_n \right\rangle$, by $\langle \phi_m |$ to give you

$$\left\langle \phi_m \left| \psi^{(1)}_{\,n} \right\rangle = \frac{\left\langle \phi_m \left| W \right| \phi_n \right\rangle}{E^{(0)}_{\,n} - E^{(0)}_{\,m}}$$

And substituting that into $\left| \psi^{(1)}_{\,n} \right\rangle = \sum_m \left\langle \phi_m \left| \psi^{(1)}_{\,n} \right\rangle \right| \phi_m \right\rangle$ gives you

$$\left| \psi^{(1)}_{\,n} \right\rangle = \sum_{m \neq n} \frac{\left\langle \phi_m \left| W \right| \phi_n \left| \phi_m \right\rangle \right.}{E^{(0)}_{\,n} - E^{(0)}_{\,m}}$$

Okay, that's your term for the first-order correction to the wave function, $| \psi^{(1)} n \rangle$. From $\left| \psi_n \right\rangle = \left| \phi_n \right\rangle + \lambda \left| \psi^{(1)}_{\,n} \right\rangle + \lambda^2 \left| \psi^{(2)}_{\,n} \right\rangle + \dots \quad (\lambda \ll 1)$, the wave function looks like this, made up of of zeroth-, first-, and second-order corrections:

$$\left| \psi_n \right\rangle = \left| \phi_n \right\rangle + \lambda \left| \psi^{(1)}_{\,n} \right\rangle + \lambda^2 \left| \psi^{(2)}_{\,n} \right\rangle + \dots \quad (\lambda \ll 1)$$

Ignoring the second-order correction for the moment and substituting $\left| \psi^{(1)}_{\,n} \right\rangle = \sum_{m \neq n} \frac{\left\langle \phi_m \left| W \right| \phi_n \left| \phi_m \right\rangle \right.}{E^{(0)}_{\,n} - E^{(0)}_{\,m}}$ in for the first-order correction gives you this for the wave function of the perturbed system, to the first order:

$$\left| \psi_n \right\rangle = \left| \phi_n \right\rangle + \sum_{m \neq n} \frac{\left\langle \phi_m \left| \lambda W \right| \phi_n \right\rangle}{E^{(0)}_{\,n} - E^{(0)}_{\,m}} \left| \phi_m \right\rangle \dots \quad (\lambda \ll 1)$$

That's the wave function of the perturbed system in terms of the perturbation. But that's still only the first-order correction. How about the second? Read on.

# Finding the second-order corrections

Now find the second-order corrections to the energy levels and the wave functions (the preceding section covers first-order corrections). To find $E^{(2)}_n$, multiply both sides of $H_0 |\psi^{(2)}_n\rangle + W |\psi^{(1)}_n\rangle = E^{(0)}_n |\psi^{(2)}_n\rangle + E^{(1)}_n |\psi^{(1)}_n\rangle + E^{(2)}_n |\phi_n\rangle$ by $\langle \phi_n |$:

$$\langle \phi_n | H_0 |\psi^{(2)}_n\rangle + \langle \phi_n | W |\psi^{(1)}_n\rangle = \langle \phi_n | E^{(0)}_n |\psi^{(2)}_n\rangle + \langle \phi_n | E^{(1)}_n |\psi^{(1)}_n\rangle + \langle \phi_n | E^{(2)}_n |\phi_n\rangle$$

This looks like a tough equation until you realize that $\langle \phi_n | \psi^{(1)}_n\rangle$ is equal to zero, so you get

$$\langle \phi_n | H_0 |\psi^{(2)}_n\rangle + \langle \phi_n | W |\psi^{(1)}_n\rangle = \langle \phi_n | E^{(0)}_n |\psi^{(2)}_n\rangle + \langle \phi_n | E^{(2)}_n |\phi_n\rangle$$

And because $\langle \phi_n | \psi^{(2)}_n\rangle$ is also equal to zero, you get

$$\langle \phi_n | W |\psi^{(1)}_n\rangle = \langle \phi_n | E^{(2)}_n |\phi_n\rangle$$

$E^{(2)}_n$ is just a number, so you have

$$|\phi_n\rangle W |\psi^{(1)}_n\rangle = E^{(2)}_n \langle \phi_n | \phi_n\rangle$$

And of course, because $\langle \phi_n | \phi_n\rangle = 1$, you have

$$E^{(2)}_n = \langle \phi_n | W |\psi^{(1)}_n\rangle$$

Note that if $|\psi^{(1)}_n\rangle$ is an eigenstate of $W$, the second-order correction equals zero.

Okay, so $E^{(2)}_n = \langle \phi_n | W | \psi^{(1)}_n\rangle$. How can you make that simpler? Well, from using $|\psi^{(1)}_n\rangle = \sum_{m \neq n} \dfrac{\langle \phi_m | W |\phi_n |\phi_m\rangle}{E^{(0)}_n - E^{(0)}_m}$. Substituting that equation into

$E^{(2)}_n = \langle \phi_n | W |\psi^{(1)}_n\rangle$ gives you

$$E^{(2)}_n = \langle \phi_n | W |\psi^{(1)}_n\rangle = \langle \phi_n | W | \sum_{m \neq n} \dfrac{\langle \phi_m | W |\phi_n\rangle}{E^{(0)}_n - E^{(0)}_m} |\phi_m\rangle$$

$$= \sum_{m \neq n} \dfrac{\left| \langle \phi_m | W |\phi_n\rangle \right|^2}{E^{(0)}_n - E^{(0)}_m}$$

Now you have $E^{(1)}{}_n = \langle \phi_n | W | \phi_n \rangle$ and $E^2{}_n = \sum_{m \neq n} \dfrac{\left| \langle \phi m | W | \phi n \rangle \right|^2}{E^{(0)}{}_n - E^{(0)}{}_m}$ . Here's the total energy with the first- and second-order corrections:

$$E_n = E^{(0)}{}_n + \lambda E^{(1)}{}_n + \lambda^2 E^{(2)}{}_n + \ldots \quad \left( \lambda \ll 1 \right)$$

So from this equation, you can say

$$E_n = E_0 + \lambda \langle \phi_n | W | \phi_n \rangle + \lambda^2 \sum_{m \neq n} \frac{\left| \langle \phi_m | W | \phi_n \rangle \right|^2}{E^{(0)}{}_n - E^{(0)}{}_m} + \ldots \quad \left( \lambda \ll 1 \right)$$

That gives you the first- and second-order corrections to the energy, according to perturbation theory.

Note that for this equation to converge, the term in the summation must be small. And note in particular what happens to the expansion term if the energy levels are degenerate:

$$\frac{\left| \langle \phi_m | W | \phi_n \rangle \right|^2}{E^{(0)}{}_n - E^{(0)}{}_m}$$

In that case, you're going to end up with an $E^{(0)}{}_n$ that equals an $E^{(0)}{}_m$, which means that the energy-corrections equation blows up, and this approach to perturbation theory is no good — which is to say that you need a different approach to perturbation theory (coming up later in "Working with Pertubations to Degenerate Hamiltonians") to handle systems with degenerate energy states.

In the next section, I show you an example to make the idea of perturbing nondegenerate Hamiltonians more real.

# Perturbation Theory to the Test: Harmonic Oscillators in Electric Fields

Consider the case in which you have a small particle oscillating in a harmonic potential, back and forth, as Figure 11-1 shows.

**Figure 11-1:**
A harmonic
oscillator.

Here's the Hamiltonian for that particle, where the particle's mass is $m$, its location is $x$, and the angular frequency of the motion is $\omega$:

$$H = \frac{-\hbar^2}{2m}\frac{d^2}{dx^2} + \frac{1}{2}m\omega^2 x^2$$

Now assume that the particle is charged, with charge $q$, and that you apply a weak electric field, $\varepsilon$, as Figure 11-2 shows.

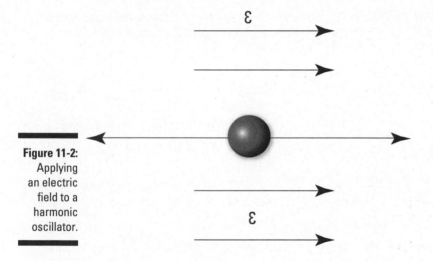

**Figure 11-2:**
Applying
an electric
field to a
harmonic
oscillator.

The force due to the electric field in this case is the perturbation, and the Hamiltonian becomes

$$H = \frac{-\hbar^2}{2m}\frac{d^2}{dx^2} + \frac{1}{2}m\omega^2 x^2 + q\varepsilon x$$

In this section, you find the energy and wave functions of the perturbed system and compare them to the exact solutions.

## Finding exact solutions

So what are the energy eigenvalues of the preceding Hamiltonian for the harmonic oscillator in an electric field? First solve for the eigenvalues exactly; then use perturbation theory. You can solve for the exact energy eigenvalues by making one of the following substitutions:

✔ $y = x + \dfrac{q\varepsilon}{m\omega^2}$

✔ $x = y - \dfrac{q\varepsilon}{m\omega^2}$

Substituting the equation solved for $x$ into $H = \dfrac{-\hbar^2}{2m}\dfrac{d^2}{dx^2} + \dfrac{1}{2}m\omega^2 x^2 + q\varepsilon x$ gives you $H = \dfrac{-\hbar^2}{2m}\dfrac{d^2}{dy^2} + \dfrac{1}{2}m\omega^2 y^2 - \dfrac{q^2\varepsilon^2}{2m\omega^2}$

The last term is a constant, so the equation is of the form

$$H = \dfrac{-\hbar^2}{2m}\dfrac{d^2}{dy^2} + \dfrac{1}{2}m\omega^2 y^2 + C$$

where $C = \dfrac{-q^2\varepsilon^2}{m\omega^2}$. $H = \dfrac{-\hbar^2}{2m}\dfrac{d^2}{dy^2} + \dfrac{1}{2}m\omega^2 y^2 + C$ is just the Hamiltonian of a harmonic oscillator with an added constant, which means that the energy levels are simply

$$E_n = \left(n + \dfrac{1}{2}\right)\hbar\omega + C$$

Substituting in for C gives you the exact energy levels:

$$E_n = \left(n + \dfrac{1}{2}\right)\hbar\omega - \dfrac{q^2\varepsilon^2}{2m\omega^2}$$

Great — that's the exact solution.

## Applying perturbation theory

As soon as you have the exact eigenvalues for your charged oscillator (see the preceding section), you have something to compare the solution from perturbation theory to. Now you can find the energy and wave functions of the perturbed system.

### Energy of the charged oscillator

So what is the energy of the charged oscillator, as given by perturbation theory? You know that the corrected energy is given by

$$E_n = E_0 + \lambda \left\langle \phi_n \middle| W \middle| \phi_n \right\rangle + \lambda^2 \sum_{m \neq n} \frac{\left| \left\langle \phi_m \middle| W \middle| \phi_n \right\rangle \right|^2}{E^{(0)}_{\ n} - E^{(0)}_{\ m}} + \dots \quad (\lambda \ll 1)$$

where $\lambda W$ is the perturbation term in the Hamiltonian. That is, here, $\lambda W = q\varepsilon x$. Now take a look at the corrected energy equation using $q\varepsilon x$ for $\lambda W$. The first-order correction is $\lambda \left\langle \phi_n \middle| W \middle| \phi_n \right\rangle$, which, using $\lambda W = q\varepsilon x$, becomes

$<\phi_n | q\varepsilon x | \phi_n>$ or $q\varepsilon <\phi_n | x | \phi_n>$

But $<\phi_n | x | \phi_n> = 0$, because that's the expectation value of $x$, and harmonic oscillators spend as much time in negative $x$ territory as in positive $x$ territory — that is, the average value of $x$ is zero. So the first-order correction to the energy, as given by perturbation theory, is zero.

Okay, what's the second-order correction to the energy, as given by perturbation theory? Here it is:

$$\lambda^2 \sum_{m \neq n} \frac{\left| \left\langle \phi_m \middle| W \middle| \phi_n \right\rangle \right|^2}{E^{(0)}_{\ n} - E^{(0)}_{\ m}} + \dots \quad (\lambda \ll 1)$$

And because $\lambda W = q\varepsilon x$, you have

$$q^2 \varepsilon^2 \sum_{m \neq n} \frac{\left| \left\langle \phi_m \middle| x \middle| \phi_n \right\rangle \right|^2}{E^{(0)}_{\ n} - E^{(0)}_{\ m}} + \dots$$

Cast this in terms of bras and kets (see Chapter 4), changing $<\phi_m|$ to $<m|$ and $|\phi_n>$ to $|n>$, making the second-order energy correction into this expression:

$$q^2 \varepsilon^2 \sum_{m \neq n} \frac{\left| \left\langle m \middle| x \middle| n \right\rangle \right|^2}{E^{(0)}_{\ n} - E^{(0)}_{\ m}} + \dots$$

You can decipher this step by step. First, the energy is

$$E^{(0)}_{\ n} = \left( n + \frac{1}{2} \right) \hbar \omega$$

That makes figuring out the second-order energy a little easier.

Also, the following expressions turn out to hold for a harmonic oscillator:

$$\blacktriangleright \langle n+1|x|n\rangle = (n+1)^{1/2} \frac{\hbar^{1/2}}{(2m\omega)^{1/2}}$$

$$\blacktriangleright \langle n-1|x|n\rangle = n^{1/2} \frac{\hbar^{1/2}}{(2m\omega)^{1/2}}$$

$$\blacktriangleright E^{(0)}_n - E^{(0)}_{n-1} = \hbar\omega$$

$$\blacktriangleright E^{(0)}_n - E^{(0)}_{n+1} = -\hbar\omega$$

With these four equations, you're ready to tackle $q^2\varepsilon^2 \sum_{m\neq n} \frac{\left|\langle m|x|n\rangle\right|^2}{E^{(0)}_n - E^{(0)}_m} + ...$, the second-order correction to the energy. Omitting higher-power terms, the summation in this equation becomes

$$q^2\varepsilon^2 \frac{\left|\langle n+1|x|n\rangle\right|^2}{E^{(0)}_n - E^{(0)}_{n+1}} + ...$$

$$q^2\varepsilon^2 \frac{\left|\langle n-1|x|n\rangle\right|^2}{E^{(0)}_n - E^{(0)}_{n-1}} ...$$

And substituting in the for $E^{(0)}_n - E^{(0)}_{n+1}$ and $E^{(0)}_n - E^{(0)}_{n-1}$ gives you

$$q^2\varepsilon^2 \frac{\left|\langle n+1|x|n\rangle\right|^2}{-\hbar\omega} + ...$$

$$q^2\varepsilon^2 \frac{\left|\langle n-1|x|n\rangle\right|^2}{\hbar\omega} ...$$

Now, substituting in for $<n + 1|x|n>$ and $<n - 1|x|n>$ gives you

$$q^2\varepsilon^2 \frac{(n+1)\hbar}{(-\hbar\omega)(2m\omega)} + ...$$

$$q^2\varepsilon^2 \frac{n\hbar}{(\hbar\omega)(2m\omega)} ...$$

or

$$-q^2\varepsilon^2 \frac{(n+1)}{2\hbar\omega^2} + ...$$

$$q^2\varepsilon^2 \frac{n}{2m\omega^2} ...$$

So the second-order correction is

$$\frac{-q^2\varepsilon^2}{2m\omega^2}\cdots$$

Therefore, according to perturbation theory, the energy of the harmonic oscillator in the electric field should be

$$E_n = \left(n+\frac{1}{2}\right)\hbar\omega - \frac{q^2\varepsilon^2}{2m\omega^2}$$

Compare this result to the earlier equation for the exact energy levels,
$$E_n = \left(n+\frac{1}{2}\right)\hbar\omega - \frac{q^2\varepsilon^2}{2m\omega^2}$$ — they're the same! In other words, perturbation theory has given you the same result as the exact answer. How's that for agreement?

Of course, you can't expect to hit the same answer every time using perturbation theory, but this result is impressive!

### Wave functions of the charged oscillator

Now figure out what the charged oscillator's wave function looks like in the presence of the electric field. Here's the wave function of the perturbed system, to the first order:

$$|\psi_n\rangle = |\phi_n\rangle + \sum_{m\neq n}\frac{\langle\phi_m|\lambda W|\phi_n\rangle}{E^{(0)}_n - E^{(0)}_m}|\phi_m\rangle\cdots \quad (\lambda \ll 1)$$

Using the <n| and |n> bras and kets you're used to for harmonic oscillators, this becomes

$$|\psi_n\rangle = |n\rangle + \sum_{m\neq n}\frac{\langle m|\lambda W|n\rangle}{E^{(0)}_n - E^{(0)}_m}|\phi_m\rangle\cdots \quad (\lambda \ll 1)$$

Because $\lambda W = q\varepsilon x$, this becomes

$$|\psi_n\rangle = |n\rangle + q\varepsilon\sum_{m\neq n}\frac{\langle m|x|n\rangle}{E^{(0)}_n - E^{(0)}_m}|m\rangle\cdots \quad (\lambda \ll 1)$$

Evidently, as with the energy, only two terms contribute, because <n|x|n> = 0. In particular, the two terms that contribute are

$$\blacktriangleright \langle n+1|x|n \rangle = (n+1)^{1/2} \frac{\hbar^{(1/2)}}{(2m\omega)^{(1/2)}}$$

$$\blacktriangleright \langle n-1|x|n \rangle = n^{1/2} \frac{\hbar^{(1/2)}}{(2m\omega)^{(1/2)}}$$

Note also that $E^{(0)}{}_n - E^{(0)}{}_{n-1} = \hbar\omega$ and $E^{(0)}{}_n - E^{(0)}{}_{n+1} = -\hbar\omega$ .

These four equations mean that

$$|\psi_n\rangle = |n\rangle + \frac{q\varepsilon}{\hbar\omega} \frac{\hbar^{1/2}}{(2m\omega)^{1/2}} \left( n^{1/2}|n-1\rangle - (n+1)^{1/2}|n+1\rangle \right)$$

Note what this equation means: Adding an electric field to a quantum harmonic oscillator spreads the wave function of the harmonic oscillator.

Originally, the harmonic oscillator's wave function is just the standard harmonic oscillator wave function, $|\psi_n\rangle$ = $|n\rangle$. Applying an electric field spreads the wave function, adding a component of $|n-1\rangle$, which is proportional to the electric field, $\varepsilon$, and the charge of the oscillator, $q$, like this:

$$|\psi_n\rangle = |n\rangle + \frac{q\varepsilon}{\hbar\omega} \frac{\hbar^{1/2}}{(2m\omega)^{1/2}} \left( n^{1/2}|n-1\rangle - \ldots \right)$$

And the wave function also spreads to the other adjacent state, $|n+1\rangle$, like this:

$$|\psi_n\rangle = |n\rangle + \frac{q\varepsilon}{\hbar\omega} \frac{\hbar^{1/2}}{(2m\omega)^{1/2}} \left( n^{1/2}|n-1\rangle - (n+1)^{1/2}|n+1\rangle \right)$$

You end up mixing states. That blending between states means that the perturbation you apply must be small with respect to the separation between unperturbed energy states, or you risk blurring the whole system to the point that you can't make any predictions about what's going to happen.

In any case, that's a nice result — blending the states in proportion to the strength of the electric field you apply — and it's typical of the result you get with perturbation theory.

Okay, that's how nondegenerate perturbation theory works. As you can see, it's strongly dependent on having the energy states separate so that your solution can blend them. But what happens when you have a system where the energies are degenerate? You take a look at that in the next section.

# Working with Perturbations to Degenerate Hamiltonians

This section tackles systems in which the energies are degenerate. Take a look at this unperturbed Hamiltonian:

$$H_0\left|\phi_{n_\alpha}\right\rangle = E^{(0)}{}_n\left|\phi_{n_\alpha}\right\rangle \quad \left(\alpha = 1,2,3\ldots\right)$$

In other words, several states have the same energy. Say the energy states are $f$-fold degenerate, like this:

$$H_0\left|\phi_{n_\alpha}\right\rangle = E^{(0)}{}_n\left|\phi_{n_\alpha}\right\rangle \quad \left(\alpha = 1,2,3,\ldots,f\right)$$

How does this affect the perturbation picture? The complete Hamiltonian, H, is made up of the original, unperturbed Hamiltonian, $H_0$, and the perturbation Hamiltonian, $H_p$:

$$H\left|\psi_n\right\rangle = \left(H_0 + H_p\left|\psi_n\right\rangle\right) = E_n\left|\psi_n\right\rangle$$

In zeroth-order approximation, you can write the eigenfunction $\left|\psi_n\right\rangle$ as a combination of the degenerate states $\left|\phi_{n_\alpha}\right\rangle$:

$$\left|\psi_n\right\rangle = \sum_{\alpha=1}^{f} a_\alpha \left|\phi_{n_\alpha}\right\rangle \ldots$$

Note that in what follows, you assume that $\langle\phi_n|\phi_n\rangle = 1$ and $\langle\phi_m|\phi_n\rangle = 0$ if $m$ is not equal to $n$. Also, you assume that the $\left|\psi_n\right\rangle$ are normalized — that is, $\langle\psi_n|\psi_n\rangle = 1$.

Plugging this zeroth-order equation into the complete Hamiltonian equation, you get

$$\sum_\alpha \left[E^{(0)}{}_n\left|\phi_{n_\alpha}\right\rangle + H_p\left|\phi_{n_\alpha}\right\rangle\right] a_\alpha = E_n \sum_\alpha a_\alpha \left|\phi_{n\alpha}\right\rangle$$

Now multiplying that equation by $\langle \phi_{n_\beta} |$ gives you

$$\sum_\alpha \left[ \left\langle \phi_{n_\beta} \left| E^{(0)}_n \right| \phi_{n_\alpha} \right\rangle + \left\langle \phi_{n_\beta} \left| H_\rho \right| \phi_{n_\alpha} \right\rangle \right] a_\alpha = E_n \sum_\alpha a_\alpha \left\langle \phi_{n_\beta} \middle| \phi_{n_\alpha} \right\rangle$$

Using the fact that $\langle \phi_n | \phi_n \rangle = 1$ and $\langle \phi_m | \phi_n \rangle = 0$ if $m$ is not equal to $n$ gives you

$$a_\beta E_n = a_\beta E^{(0)}_n + \sum_{\alpha=1}^f a_\alpha \left\langle \phi_{n_\beta} \left| H_\rho \right| \phi_{n_\alpha} \right\rangle$$

Physicists often write that equation as

$$\sum_{\alpha=1}^f a_\alpha H_{\rho_{\alpha\beta}} - \left( a_\beta E_n - a_\beta E^{(0)}_n \right) = 0 \quad \left( \beta = 1,2,3,\ldots,f \right)$$

where $H_{\rho_{\alpha\beta}} = \langle \phi_{n_\beta} | H_\rho | \phi_{n_\alpha} \rangle$. And people also write that equation as

$$\sum_{\alpha=1}^f a_\alpha H_{\rho_{\alpha\beta}} - a_\beta E^{(1)}_n = 0 \quad \left( \beta = 1,2,3,\ldots,f \right)$$

where $E^{(1)}_n = E_n - E^{(0)}_n$. That's a system of linear equations, and the solution exists only when the determinant to this array is nonvanishing:

$$\begin{vmatrix} H_{\rho 11} - E^{(1)}_n & H_{\rho 12} & H_{\rho 13} & \cdots & H_{\rho 1f} \\ H_{\rho 21} & H_{\rho 22} - E^{(1)}_n & H_{\rho 23} & \cdots & H_{\rho 1f} \\ & \cdot & & & \\ & \cdot & & & \\ & \cdot & & & \\ H_{\rho f1} & H_{\rho f2} & H_{\rho f3} & \cdots & H_{\rho ff} - E^{(1)}_n \end{vmatrix}$$

The determinant of this array is an $f$th degree equation in $E^{(1)}_n$, and it has $f$ different roots, $E^{(1)}_{n_\alpha}$. Those $f$ different roots are the first-order corrections to the Hamiltonian. Usually, those roots are different because of the applied perturbation. In other words, the perturbation typically gets rid of the degeneracy.

So here's the way you find the eigenvalues to the first order — you set up an $f$-by-$f$ matrix of the perturbation Hamiltonian, $H_p$, where $H_{p_{\alpha\beta}} = \langle \phi_{n_\beta} | H_p | \phi_{n_\beta} \rangle$:

$$\begin{vmatrix} H_{\rho 11} & H_{\rho 12} & H_{\rho 13} & \cdots & H_{\rho 1f} \\ H_{\rho 21} & H_{\rho 22} & H_{\rho 23} & \cdots & H_{\rho 2f} \\ & & \cdot & & \\ & & \cdot & & \\ & & \cdot & & \\ H_{\rho f1} & H_{\rho f2} & H_{\rho f3} & \cdots & H_{\rho ff} \end{vmatrix}$$

Then diagonalize this matrix and determine the $f$ eigenvalues $E^{(1)}{}_{n_\alpha}$ and the matching eigenvectors:

$$\begin{vmatrix} a_{\alpha 1} \\ a_{\alpha 2} \\ a_{\alpha 3} \\ \cdot \\ \cdot \\ \cdot \\ a_{\alpha f} \end{vmatrix} = a_\beta \qquad \beta = 1, 2, 3, \ldots f$$

Then you get the energy eigenvalues to first order this way:

$$E_{n_\alpha} = E^{(0)}{}_n + E^{(1)}{}_{n_\alpha} \qquad \left( \alpha = 1, 2, 3, \ldots, f \right)$$

And the eigenvectors are

$$\left| \psi_{n_\alpha} \right\rangle = \sum_{\beta=1}^{f} a_{\alpha\beta} \left| \phi_{n_\beta} \right\rangle$$

In the next section, you look at an example to clarify this idea.

# Testing Degenerate Perturbation Theory: Hydrogen in Electric Fields

In this section, you see whether degenerate perturbation theory can handle the hydrogen atom, which has energy states degenerate in different angular momentum quantum numbers, when you remove that degeneracy by applying an electric field. This setup is called the *Stark effect*.

Specifically, suppose you apply an electric field, ε, to a hydrogen atom in the $n = 2$ excited state. That state has four eigenfunctions that have the same energy, where the quantum numbers are $|nlm>$ (note that you're renaming these eigenfunctions $|1>$, $|2>$, and so on to make the calculation easier):

- ✔ $|1> = |200>$
- ✔ $|2> = |211>$
- ✔ $|3> = |210>$
- ✔ $|4> = |21-1>$

All these unperturbed states have the same energy, $E = -R/4$, where R is the *Rydberg constant,* 13.6 eV. But at least some of these states will have their energies changed when you apply the electric field.

What does the electric field, ε, cause the perturbation Hamiltonian, $H_p$, to become? Here's the perturbation Hamiltonian:

$$H_p = e\varepsilon z$$

So you have to evaluate this equation for the various states. For example, what is the following expression equal to, where $<1| = <200|$ and $|3> = |210>$?

$$<1|H_p|3>$$

You solve for the unperturbed hydrogen wave functions in Chapter 9. In general, here's what the wave function $\psi_{nlm}(r, \theta, \phi)$ looks like for hydrogen:

$$\psi_{nlm}(r,\theta,\phi) = \frac{\left(\frac{2}{nr_0}\right)^{3/2}\left[(n-l-1)!\right]^{1/2}}{\left[2n(n+1)!\right]^{1/2}} e^{-r/nr_0}\left(\frac{2r}{nr_0}\right)^l L_{n-l-1}^{2l+1}\left(\frac{2r}{nr_0}\right) Y_{lm}(\theta,\phi)$$

where $L_{n-l-1}^{2l+1}(2r/nr_0)$ is a generalized Laguerre polynomial. Doing all the math gives you the following result, where $a_0$ is the Bohr radius of the atom:

$$\left\langle 1\left|H_p\right|3\right\rangle = e\varepsilon\left\langle 1\left|z\right|3\right\rangle = -3e\varepsilon a_0$$

The $<1|H_p|3>$ is just one term you have to compute, of course. Here's the full matrix for the perturbation Hamiltonian connecting all states, where $H_{p_{\alpha\beta}} = <\alpha|H_p|\beta>$:

$$\begin{vmatrix} H_{\rho 11} & H_{\rho 12} & H_{\rho 13} & H_{\rho 14} \\ H_{\rho 21} & H_{\rho 22} & H_{\rho 23} & H_{\rho 24} \\ H_{\rho 31} & H_{\rho 32} & H_{\rho 33} & H_{\rho 34} \\ H_{\rho 41} & H_{\rho 42} & H_{\rho 43} & H_{\rho 44} \end{vmatrix}$$

Doing the math gives you this remarkably simple result:

$$\begin{vmatrix} H_{\rho 11} & H_{\rho 12} & H_{\rho 13} & H_{\rho 14} \\ H_{\rho 21} & H_{\rho 22} & H_{\rho 23} & H_{\rho 24} \\ H_{\rho 31} & H_{\rho 32} & H_{\rho 33} & H_{\rho 34} \\ H_{\rho 41} & H_{\rho 42} & H_{\rho 43} & H_{\rho 44} \end{vmatrix} = -3e\varepsilon a_0 \begin{vmatrix} 0 & 0 & 1 & 0 \\ 0 & 0 & 0 & 0 \\ 1 & 0 & 0 & 0 \\ 0 & 0 & 0 & 0 \end{vmatrix}$$

Diagonalizing this matrix gives you these eigenvalues — the first-order corrections to the unperturbed energies:

✔ $E^{(1)}_1 = -3e\varepsilon a_0$

✔ $E^{(1)}_2 = 0$

✔ $E^{(1)}_3 = 3e\varepsilon a_0$

✔ $E^{(1)}_4 = 0$

where $E^{(1)}_1$ is the first-order correction to the energy of the $|1\rangle$ eigenfunction, $E^{(1)}_2$ is the first-order correction to the energy of the $|2\rangle$ eigenfunction, and so on. Adding these corrections to the unperturbed energy for the $n = 2$ state gives you the final energy levels:

✔ $E_1 = \dfrac{-R}{4} - 3e\varepsilon a_0$

✔ $E_2 = \dfrac{-R}{4}$

✔ $E_3 = \dfrac{-R}{4} + 3e\varepsilon a_0$

✔ $E_4 = \dfrac{-R}{4}$

where R is the Rydberg constant. Note this result: The Stark effect removes the energy degeneracy in $|200\rangle$ and $|210\rangle$ (the $|1\rangle$ and $|3\rangle$ eigenfunctions), but the degeneracy in $|211\rangle$ and $|21-1\rangle$ (the $|2\rangle$ and $|4\rangle$ eigenfunctions) remains.

# Chapter 12

# Wham-Blam! Scattering Theory

. . . . . . . . . . . . . . . . . . . . . . . . . . . . . . . . . . . . . . . . . .

## In This Chapter

▶ Switching between lab and center-of-mass frames

▶ Solving the Schrödinger equation

▶ Finding the wave function

▶ Putting the Born approximation to work

. . . . . . . . . . . . . . . . . . . . . . . . . . . . . . . . . . . . . . . . . .

*Y*our National Science Foundation grant finally came through, and you built your new synchrotron — a particle accelerator. Electrons and anti-electrons accelerate at near the speed of light along a giant circular track enclosed in a vacuum chamber and collide, letting you probe the structure of the high-energy particles you create. You're sitting at the console of your giant new experiment, watching the lights flashing and the signals on the screens approvingly. Millions of watts of power course through the thick cables, and the radiation monitors are beeping, indicating that things are working. Cool.

You're accelerating particles and smashing them against each other to observe how they scatter. But this is slightly more complex than observing how pool balls collide. Classically, you can predict the exact angle at which colliding objects will bounce off each other if the collision is *elastic* (that is, momentum and kinetic energy are both conserved). Quantum mechanically, however, you can only assign probabilities to the angles at which things scatter.

Physicists use large particle accelerators to discover more about the structure of matter, and that study is central to modern physics. This chapter serves as an introduction to that field of study. You get to take a look at particle scattering on the subatomic level.

# Introducing Particle Scattering and Cross Sections

Think of a scattering experiment in terms of *particles in* and *particles out*. Look at Figure 12-1, for example. In the figure, particles are being sent in a stream from the left and interacting with a target; most of them continue on unscattered, but some particles interact with the target and scatter.

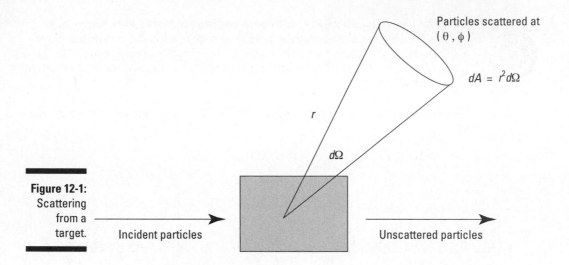

Particles scattered at
$(\theta, \phi)$

$dA = r^2 d\Omega$

$r$

$d\Omega$

**Figure 12-1:**
Scattering
from a
target.

Incident particles

Unscattered particles

Those particles that do scatter do so at a particular angle in three dimensions — that is, you give the scattering angle as a solid angle, $d\Omega$, which equals $\sin\theta\, d\theta\, d\phi$, where $\phi$ and $\theta$ are the spherical angles I introduce in Chapter 8.

The number of particles scattered into a specific $d\Omega$ per unit time is proportional to a very important quantity in scattering theory: the differential cross section.

The *differential cross section* is given by $\dfrac{d\sigma(\phi,\theta)}{d\Omega}$, and it's a measure of the number of particles per second scattered into $d\Omega$ per incoming flux. The *incident flux*, J (also called the *current density*), is the number of incident particles per unit area per unit time. So $\dfrac{d\sigma(\phi,\theta)}{d\Omega}$ is

$$\frac{d\sigma(\phi,\theta)}{d\Omega} = \frac{1}{J}\frac{dN(\phi,\theta)}{d\Omega}$$

where $N(\phi, \theta)$ is the number of particles at angles $\phi$ and $\theta$.

The differential cross section $\dfrac{d\sigma(\phi,\theta)}{d\Omega}$ has the dimensions of area, so calling it a cross section is appropriate. The cross section is sort of like the size of the bull's eye when you're aiming to scatter incident particles through a specific solid angle.

The *differential cross section* is the cross section for scattering to a specific solid angle. The *total cross section,* σ, is the cross section for scattering of any kind, through any angle. So if the differential cross section for scattering to a particular solid angle is like the bull's eye, the total cross section corresponds to the whole target.

You can relate the total cross section to the differential cross section by integrating the following:

$$\sigma = \int \frac{d\sigma(\phi,\theta)d\Omega}{d\Omega} = \int_0^\pi \sin\theta d\theta \int_0^{2\pi} \frac{d\sigma(\phi,\theta)d\varphi}{d\Omega}$$

# Translating between the Center-of-Mass and Lab Frames

Now you can start getting into the details of scattering, beginning with a discussion of the center-of-mass frame versus the lab frame. Experiments take place in the *lab frame,* but you do scattering calculations in the *center-of-mass frame,* so you have to know how to translate between the two frames. This section explains how the frames differ and shows you how to relate the scattering angles and cross sections when you change frames.

## Framing the scattering discussion

Look at Figure 12-2 — that's scattering in the lab frame. One particle, traveling at $v_{1lab}$, is incident on another particle that's at rest ($v_{2lab} = 0$) and hits it. After the collision, the first particle is scattered at angle $\theta_1$, traveling at $v_1 lab$, and the other particle is scattered at angle $\theta_2$ and velocity $v_{2lab}$.

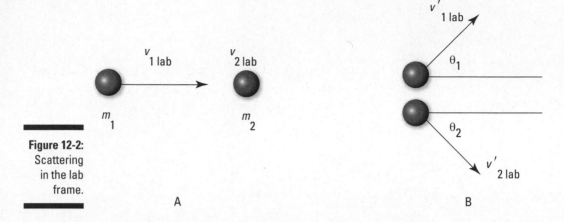

**Figure 12-2:**
Scattering
in the lab
frame.

Now in the center-of-mass frame, the center of mass is stationary and the particles head toward each other. After they collide, they head away from each other at angles θ and π – θ.

You have to move back and forth between these two frames — the lab frame and the center-of-mass frame — so you need to relate the velocities and angles (in a nonrelativistic way).

## Relating the scattering angles between frames

To relate the angles $\theta_1$ and $\theta$, you start by noting that you can connect $v_{1lab}$ and $v_{1c}$ using the velocity of the center of mass, $v_{cm}$, this way:

$$\boldsymbol{v}_{1lab} = \boldsymbol{v}_{1c} + \boldsymbol{v}_{cm}$$

In addition, here's what can say about the velocity of particle 1 after it collides with particle 2:

$$\boldsymbol{v'}_{1lab} = \boldsymbol{v'}_{1c} + \boldsymbol{v}_{cm}$$

Now you can find the components of these velocities:

✔ $\boldsymbol{v'}_{1lab} \cos\theta_1 = \boldsymbol{v'}_{1c} \cos\theta + \boldsymbol{v}_{cm}$

✔ $\boldsymbol{v'}_{1lab} \sin\theta_1 = \boldsymbol{v'}_{1c} \sin\theta$

Dividing the equation in the second bullet by the one in the first gives you

$$\tan\theta_1 = \frac{\sin\theta}{\cos\theta + v_{cm}/v'_{1c}}$$

But wouldn't it be easier if you could relate $\theta_1$ and $\theta$ by something that didn't involve the velocities, only the masses, such as the following?

$$\tan\theta_1 = \frac{\sin\theta}{\cos\theta + m_1/m_2}$$

Well, you can. To see that, start with

$$v_{cm} = \frac{m_1}{m_1 + m_2} v_{1lab}$$

And you can show that

$$v_{1c} = \frac{m_2}{m_1 + m_2} v_{1lab}$$

You can also use the conservation of momentum to say what happens after the collision. In fact, because the center of mass is stationary in the center-of-mass frame, the total momentum before and after the collision is zero in that frame, like this:

$$m_1 v_{1c} - m_2 v_{2c} = 0$$

Therefore

$$v_{2c} = \frac{m_1}{m_2} v_{1c}$$

And after the collision, $m_1 v'_{1c} \cos\theta - m_2 v'_{2c} \cos\theta = 0$, which means that

$$v'_{2c} = \frac{m_1}{m_2} v'_{1c}$$

Also, if the collision is elastic (and you assume all collisions are elastic in this chapter), kinetic energy is conserved in addition to momentum, so that means the following is true:

$$\frac{1}{2} m_1 v_{1c}^2 + \frac{1}{2} m_2 v_{2c}^2 = \frac{1}{2} m_1 v'_{1c}^2 + \frac{1}{2} m_2 v'_{2c}^2$$

Substituting $v_{2c} = \dfrac{m_1}{m_2} v_{1c}$ and $v'_{2c} = \dfrac{m_1}{m_2} v'_{1c}$ into this equation gives you

$$v'_{1c} = v_{1c}$$
$$\text{and } v'_{2c} = v_{2c}$$

Given these two equations, you can redo $v_{1c} = \dfrac{m_2}{m_1 + m_2} v_{1lab}$ as

$$v'_{1c} = v_{1c} = \frac{m_2}{m_1 + m_2} v_{1lab}$$

Dividing $v_{cm} = \dfrac{m_1}{m_1 + m_2} v_{1lab}$ by that equation gives you

$$\frac{v_{cm}}{v'_{1c}} = \frac{m_1}{m_2}$$

And because you saw earlier that $\tan\theta_1 = \dfrac{\sin\theta}{\cos\theta + v_{cm}/v'_{1c}}$ , substituting

$\dfrac{v_{cm}}{v'_{1c}} = \dfrac{m_1}{m_2}$ into this equation gives you at last

$$\tan\theta_1 = \frac{\sin\theta}{\cos\theta + m_1/m_2}$$

Okay, that relates $\theta_1$ and $\theta$, which is what you were trying to do. Using the relation $\cos\theta_1 = \dfrac{1}{\left(\tan^2\theta_1 + 1\right)^{1/2}}$ , you can rewrite $\tan\theta_1 = \dfrac{\sin\theta}{\cos\theta + m_1/m_2}$ as the following:

$$\cos\theta_1 = \frac{\cos\theta + m_1/m_2}{\left[ 1 + m_1^2/m_2^2 + 2\cos\theta\left(m_1/m_2\right) \right]^{1/2}}$$

You can also relate $\theta_2$ and $\theta$. You can show that $\tan\theta_2 = \cot\left(\theta/2\right)$, which, using a little trig, means that

$$\theta_2 = \frac{\pi - \theta}{2}$$

Okay, now you've related the angles between the lab and center-of-mass frames. How about relating the cross sections in the two frames? That's in the next section.

# Translating cross sections between the frames

The preceding section relates $\theta_1$ and $\theta$ and $\theta_2$ — the angles of the scattered particles in the lab frame and the center-of-mass frame. Now how about relating the differential cross section — the bull's eye when you're aiming to scatter the particles at a particular angle — between the lab and center-of-mass frames?

The differential $d\sigma$ (total cross section) is infinitesimal in size, and it stays the same between the two frames. But the angles that make up $d\Omega$, the scattering angle, vary when you translate between frames. You get to take a look at how that works now, relating the lab differential cross section:

$$\left.\frac{d\sigma(\phi,\theta)}{d\Omega}\right|_{lab}$$

to the center-of-mass differential cross section:

$$\left.\frac{d\sigma(\phi,\theta)}{d\Omega}\right|_{cm}$$

In the lab frame, $d\Omega_1 = \sin\theta_1\, d\theta_1\, d\phi_1$. And in the center-of-mass frame, $d\Omega = \sin\theta\, d\theta\, d\phi$. Because $d\sigma_{lab} = d\sigma_{cm}$, the following equation is true:

$$\left.\frac{d\sigma(\phi,\theta)}{d\Omega_1}\right|_{lab}^{d\Omega_1} = \left.\frac{d\sigma(\phi,\theta)}{d\Omega}\right|_{cm}^{d\Omega}$$

Putting that equation with the equations for the lab frame and the center-of-mass frame, you have

$$\left.\frac{d\sigma(\phi,\theta)}{d\Omega_1}\right|_{lab} = \left.\frac{d\sigma(\phi,\theta)}{d\Omega}\right|_{cm}\frac{\sin\theta}{\sin\theta_1}\frac{d\theta}{d\theta_1}\frac{d\phi}{d\phi_1}$$

Because you have cylindrical symmetry here, $\phi = \phi_1$, so

$$\left.\frac{d\sigma(\phi,\theta)}{d\Omega_1}\right|_{lab} = \left.\frac{d\sigma(\phi,\theta)}{d\Omega}\right|_{cm}\frac{\sin\theta}{\sin\theta_1}\frac{d\theta}{d\theta_1} = \left.\frac{d\sigma(\phi,\theta)}{d\Omega}\right|_{cm}\frac{d(\cos\theta)}{d(\cos\theta_1)}$$

You've already seen that $\cos\theta_1 = \dfrac{\cos\theta + m_1/m_2}{\left[1 + m_1^2/m_2^2 + 2\cos\theta\left(m_1/m_2\right)\right]^{1/2}}$ , so

$$\frac{d\left(\cos\theta_1\right)}{d\left(\cos\theta\right)} = \frac{1 + \cos\theta\left(m_1/m_2\right)}{\left[1 + m_1^2/m_2^2 + 2\cos\theta\left(m_1/m_2\right)\right]^{3/2}}$$

. Therefore

$$\left.\frac{d\sigma(\phi,\theta)}{d\Omega_1}\right|_{lab} = \frac{\left[1 + m_1^2/m_2^2 + 2\cos\theta\left(m_1/m_2\right)\right]^{3/2}}{1 + \cos\theta\left(m_1/m_2\right)} \left.\frac{d\sigma(\phi,\theta)}{d\Omega}\right|_{cm}$$

You can also show that

$$\left.\frac{d\sigma(\phi,\theta)}{d\Omega_2}\right|_{lab} = 4\sin\left(\frac{\theta}{2}\right)\left.\frac{d\sigma(\phi,\theta)}{d\Omega}\right|_{cm}$$

## Trying a lab-frame example with particles of equal mass

Say you have two particles of equal mass colliding in the lab frame (where one particle starts at rest). You want to show that the two particles end up traveling at right angles with respect to each other in the lab frame.

Note that if $m_1 = m_2$, then $\cos\theta_1 = \dfrac{\cos\theta + m_1/m_2}{\left[1 + m_1^2/m_2^2 + 2\cos\theta\left(m_1/m_2\right)\right]^{1/2}}$ gives $\tan(\theta_1) =$

$\tan(\theta/2)$, so $\theta_1 = \theta/2$. And $\left.\dfrac{d\sigma(\phi,\theta)}{d\Omega_1}\right|_{lab} = \dfrac{\left[1 + m_1^2/m_2^2 + 2\cos\theta\left(m_1/m_2\right)\right]^{3/2}}{1 + \cos\theta\left(m_1/m_2\right)} \left.\dfrac{d\sigma(\phi,\theta)}{d\Omega}\right|_{cm}$

becomes

$$\left.\frac{d\sigma(\phi,\theta)}{d\Omega_1}\right|_{lab} = 4\cos\left(\frac{\theta}{2}\right)\left.\frac{d\sigma(\phi,\theta)}{d\Omega}\right|_{cm}$$

Note also that $\tan(\theta_2) = \cot(^\theta/_2)$, or $\tan(\theta_2) = \tan(^\pi/_2 - ^\theta/_2)$.

You know that $\theta_1 = ^\theta/_2$, and $\tan(\theta_2) = \tan(^\pi/_2 - ^\theta/_2)$ tells you that the following is true:

$$\theta_2 = ^\pi/_2 - ^\theta/_2$$

So substituting $\theta_1 = ^\theta/_2$ into the preceding equation gives you

$$\theta_2 = ^\pi/_2 - \theta_1$$
$$\theta_2 + \theta_1 = ^\pi/_2$$

Therefore, $\theta_2$ and $\theta_1$, the angles of the particles in the lab frame after the collision, add up to $^\pi/_2$ — which means $\theta_2$ and $\theta_1$ are at right angles with respect to each other. Cool.

In this case, you can use the relations you've already derived to get these relations in the special case where $m_1 = m_2$:

$$\left.\frac{d\sigma(\phi,\theta)}{d\Omega_1}\right|_{lab} = 4\cos(\theta_1)\left.\frac{d\sigma(\phi,\theta)}{d\Omega}\right|_{cm}$$

$$\left.\frac{d\sigma(\phi,\theta)}{d\Omega_1}\right|_{lab} = 4\cos\left(\frac{\theta}{2}\right)\left.\frac{d\sigma(\phi,\theta)}{d\Omega}\right|_{cm}$$

$$\left.\frac{d\sigma(\phi,\theta)}{d\Omega_2}\right|_{lab} = 4\cos(\theta_2)\left.\frac{d\sigma(\phi,\theta)}{d\Omega}\right|_{cm}$$

$$\left.\frac{d\sigma(\phi,\theta)}{d\Omega_2}\right|_{lab} = 4\cos\left(\frac{\theta}{2}\right)\left.\frac{d\sigma(\phi,\theta)}{d\Omega}\right|_{cm}$$

# Tracking the Scattering Amplitude Of Spinless Particles

In the earlier section "Translating between the Center-of-Mass and Lab Frames," you see how to translate from the lab frame to the center-of-mass frame and back again, and those translations work classically as well as in quantum physics (as long as the speeds involved are nonrelativistic). Now you look at the elastic scattering of two spinless nonrelativistic particles from the time-independent quantum physics point of view.

Assume that the interaction between the particles depends only on their relative distance, $|r_1 - r_2|$. You can reduce problems of this kind to two decoupled problems (see Chapter 9 for details). The first decoupled equation treats the center of mass of the two particles as a free particle, and the second equation is for a fictitious particles of mass $\dfrac{m_1 m_2}{m_1 + m_2}$.

The first decoupled equation, the free-particle equation of the center of mass, is of no interest to you in scattering discussions. The second equation is the one to concentrate on, where $\mu = \dfrac{m_1 m_2}{m_1 + m_2}$:

$$\frac{-\hbar^2}{2\mu}\nabla^2\psi(r)+V(r)\psi(r)=E\psi(r)$$

You can use the preceding equation to solve for the probability that a particle is scattered into a solid angle $d\Omega$ — and you give this probability by the differential cross section, $\dfrac{d\sigma}{d\Omega}$.

In quantum physics, wave packets represent particles. In terms of scattering, these wave packets must be wide enough so that the spreading that occurs during the scattering process is negligible (however, the wave packet can't be so spread that it encompasses the whole lab, including the particle detectors). Here's the crux: After the scattering, the wave function breaks up into two parts — an unscattered part and a scattered part. That's how scattering works in the quantum physics world.

## The incident wave function

Assume that the scattering potential $V(r)$ has a very finite range, $a$. Outside that range, the wave functions involved act like free particles. So the incident particle's wave function, outside the limit of $V(r)$ — that is, outside the range $a$ from the other particle — is given by this equation, because $V(r)$ is zero:

$$\nabla^2\phi_{inc}(r)+k_0\phi_{inc}(r)=0$$

where $k_0 = \dfrac{2\mu E_0}{\hbar^2}$.

The form $\nabla^2\phi_{inc}(r)+k_0\phi_{inc}(r)=0$ is the equation for a plane wave, so $\phi_{inc}(r)$ is $\phi_{inc}(r) = Ae^{ik_0 \cdot r}$, where A is a normalization factor and $\boldsymbol{k}_0 \cdot \boldsymbol{r}$ is the dot product between the incident wave's wave vector and $\boldsymbol{r}$. In other words, you're treating the incident particle as a particle of momentum $P_0 = \hbar k_0$.

# The scattered wave function

After the scattering of the spinless particles, the nonscattered wave function isn't of much interest to you, but the scattered wave function is. Although the incident wave function has the form $\phi_{inc}(r) = Ae^{ik_0 \cdot r}$, the scattered wave function has a slightly different form:

$$\phi_{sc}(r) = A\mathrm{f}(\phi,\theta)\frac{e^{ik \cdot r}}{r}$$

The f($\phi$, $\theta$) part is called the *scattering amplitude*, and your job is to find it. Here, A is a normalization factor and

$$|k| = \frac{2\mu E}{\hbar^2}$$

where E is the energy of the scattered particle.

# Relating the scattering amplitude and differential cross section

The scattering amplitude of spinless particles turns out to be crucial to understanding scattering from the quantum physics point of view. To see that, take a look at the flux densities, $J_{inc}$ (the flux density of the incident particle) and $J_{sc}$ (the flux density for the scattered particle):

- $J_{inc} = \frac{i\hbar}{2\mu}\left(\phi_{inc}\nabla\phi^*_{inc} - \phi^*_{inc}\nabla\phi_{inc}\right)$

- $J_{sc} = \frac{i\hbar}{2\mu}\left(\phi_{sc}\nabla\phi^*_{sc} - \phi^*_{sc}\nabla\phi_{sc}\right)$

Inserting your expressions for $\phi_{inc}$ and $\phi_{sc}$ into these equations gives you the following, where $f(\phi, \theta)$ is the scattering amplitude:

- $J_{inc} = |A|^2 \frac{\hbar k_0}{\mu}$

- $J_{sc} = |A|^2 \frac{\hbar k}{\mu r^2}|\mathrm{f}(\phi,\theta)|^2$

Now in terms of the flux, the number of particles $dN(\phi, \theta)$ scattered into $d\Omega$ and passing through an area $dA = r^2 d\Omega$ is

$$dN(\phi, \theta) = J_{sc}r^2 d\Omega$$

Plugging in $J_{sc} = |A|^2 \dfrac{\hbar k}{\mu r^2} |f(\phi,\theta)|^2$ into the preceding equation gives you

$$\frac{dN(\phi,\theta)}{d\Omega} = |A|^2 \frac{\hbar k}{\mu} |f(\phi,\theta)|^2$$

Also, recall from the beginning of the chapter that $\dfrac{d\sigma(\phi,\theta)}{d\Omega} = \dfrac{1}{J}\dfrac{dN(\phi,\theta)}{d\Omega}$ . You get

$$\frac{d\sigma(\phi,\theta)}{d\Omega} = \frac{1}{J_{inc}}\frac{dN(\phi,\theta)}{d\Omega} = \frac{k}{k_0}|f(\phi,\theta)|^2$$

And here's the trick — for elastic scattering, $k = k_0$, which means that this is your final result:

$$\frac{d\sigma(\phi,\theta)}{d\Omega} = |f(\phi,\theta)|^2$$

The problem of determining the differential cross section breaks down to determining the scattering amplitude.

# Finding the scattering amplitude

To find the scattering amplitude — and therefore the differential cross section — of spinless particles, you work on solving the Schrödinger equation: $\dfrac{-\hbar^2}{2\mu}\nabla^2\psi(r) + V(r)\psi(r) = E\psi(r)$. You can also write this as

$$(\nabla^2 + k^2)\psi(r) = \frac{2\mu}{\hbar^2} + V(r)\psi(r)$$

You can express the solution to that differential equation as the sum of a homogeneous solution and a particular solution:

$$\psi(r) = \psi_h(r) + \psi_p(r)$$

The homogeneous solution satisfies this equation:

$$(\nabla^2 + k^2)\psi_h(r) = \frac{2\mu}{\hbar^2}V(r)\psi_h(r)$$

And the homogeneous solution is a plane wave — that is, it corresponds to the incident plane wave:

$Ae^{ik_0 \cdot r}$

To take a look at the scattering that happens, you have to find the particular solution. You can do that in terms of *Green's functions*, so the solution to $\left(\nabla^2 + k^2\right)\psi(r) = \dfrac{2\mu}{\hbar^2} + V(r)\psi(r)$ is

$$\psi(r) = Ae^{k_0 \cdot r} + \frac{2\mu}{\hbar^2}\int G(r - r')V(r')\psi(r')d^3r'$$

where $G(r - r') = \dfrac{1}{(2\pi)^3}\int \dfrac{e^{iq(r-r')}}{k^2 - q^2}d^3q$ .

This integral breaks down to

$$G(r - r') = \frac{-1}{4\pi^2 i |r - r'|}\int \frac{qe^{iq|r-r'|}}{q^2 - k^2}dq$$

You can solve the preceding equation in terms of incoming and/or outgoing waves. Because the scattered particle is an outgoing wave, the Green's function takes this form:

$$G(r - r') = \frac{-e^{ik|r-r'|}}{4\pi |r - r'|}$$

You already know that

$$\psi(r) = Ae^{ik_0 \cdot r} + \frac{2\mu}{\hbar^2}\int G(r - r')V(r')\psi(r')d^3r'$$

So substituting $G(r - r') = \dfrac{-e^{ik|r-r'|}}{4\pi |r - r'|}$ into the preceding equation gives you

$$\psi(r) = Ae^{ik_0 \cdot r} - \frac{\mu}{2\pi\hbar^2}\int \frac{e^{ik|r\cdot r'|}}{|r - r'|}V(r')\psi(r')d^3r'$$

Wow, that's an integral equation for $\psi(r)$ , the wave equation — how do you go about solving this whopper? Why, you use the Born approximation, of course.

# The Born Approximation: Rescuing the Wave Equation

Okay, your dilemma is to solve the following equation for $\psi(r)$, where $\phi_{inc} = Ae^{ik_0r}$:

$$\psi(r) = \phi_{inc} - \frac{\mu}{2\pi\hbar^2} \int \frac{e^{ik|r-r'|}}{|r-r'|} V(r')\psi(r')d^3r'$$

You can do that with a series of successive approximations, called the *Born approximation* (this is a famous result). To start, the zeroth order Born approximation is just $\psi_0(r) = \phi_{inc}(r)$. And substituting this zeroth-order term, $\psi_0(r)$, into the first equation in this section gives you the first-order term:

$$\psi_1(r) = \phi_{inc} - \frac{\mu}{2\pi\hbar^2} \int \frac{e^{ik|r-r_1|}}{|r-r_1|} V(r_1)\psi_0(r_1)d^3r_1$$

which, using $\psi_0(r) = \phi_{inc}(r)$ gives you

$$\psi_1(r) = \phi_{inc} - \frac{\mu}{2\pi\hbar^2} \int \frac{e^{ik|r-r_1|}}{|r-r_1|} V(r_1)\phi_{inc}(r_1)d^3r_1$$

You get the second-order term by substituting this equation into $\psi(r) = \phi_{inc} - \frac{\mu}{2\pi\hbar^2} \int \frac{e^{ik|r-r'|}}{|r-r'|} V(r')\psi(r')d^3r'$ :

$$\psi_2(r) = \phi_{inc} - \frac{\mu}{2\pi\hbar^2} \int \frac{e^{ik|r-r_2|}}{|r-r_2|} V(r_2)\psi_1(r_2)d^3r_2$$

And substituting $\psi_1(r) = \phi_{inc} - \frac{\mu}{2\pi\hbar^2} \int \frac{e^{ik|r-r_1|}}{|r-r_1|} V(r_1)\psi_0(r_1)d^3r_1$ into the preceding equation gives you

$$\psi_2(r) =$$

$$\phi_{inc} - \frac{\mu}{2\pi\hbar^2} \int \frac{e^{ik|r-r_2|}}{|r-r_2|} V(r_2)\phi_{inc}(r_2)d^3r_2$$

$$+ \frac{\mu^2}{4\pi^2\hbar^4} \int \frac{e^{ik|r-r_2|}}{|r-r_2|} V(r_2)\psi_1(r_2)d^3r_2 \int \frac{e^{ik|r_2-r_1|}}{|r_2-r_1|} V(r_1)\phi_{inc}(r_1)d^3r_1$$

The pattern continues for the higher terms, which you can find by plugging lower-order terms into higher ones.

## Exploring the far limits of the wave function

Now that you've used the Born approximation (see the preceding section), take a look at the case where $r$ is large — in scattering experiments, $r \gg r'$, where $r$ is the distance from the target to the detector and $r'$ is the size of the detector. What happens to $\psi(r) = \phi_{inc} - \dfrac{\mu}{2\pi\hbar^2}\int \dfrac{e^{ik|r-r'|}}{|r-r'|}V(r')\psi(r')d^3r'$ , the exact integral equation for the wave function, when $r \gg r'$? Here's the answer:

$$\psi(r) = \phi_{inc} + \frac{\mu}{2\pi\hbar^2}\int \frac{e^{ik|r-r'|}}{|r-r|}V(r')\psi(r')d^3r'$$

Because $r \gg r'$, you can say that $k|r-r'| \approx kr - \boldsymbol{k}\cdot\boldsymbol{r'}$, where $\boldsymbol{k}\cdot\boldsymbol{r'}$ is the dot product of $\boldsymbol{k}$ and $\boldsymbol{r'}$ ($\boldsymbol{k}$ is the wave vector of the scattered particle). And

$$\frac{1}{|r-r'|} \approx \frac{1}{r}$$

Using the last two equations in $\psi(r) = \phi_{inc} + \dfrac{\mu}{2\pi\hbar^2}\int \dfrac{e^{ik|r-r'|}}{|r-r|}V(r')\psi(r')d^3r'$ gives you

$$\psi(r) = Ae^{ik_0\cdot r} + \frac{Ae^{ikr}}{r}f(\phi,\theta) \qquad r \to \infty$$

And here

$$f(\theta,\phi) = \frac{-\mu_2}{2\pi\hbar^2}\int e^{-ik\cdot r'}V(r')\psi(r')d^3r' = \frac{-\mu_2}{2\pi\hbar^2}\langle\phi|V|\psi\rangle$$

The differential cross section is given by $\dfrac{d\sigma(\phi,\theta)}{d\Omega} = |f(\phi,\theta)|^2$ , which in this case becomes

$$\frac{d\sigma(\phi,\theta)}{d\Omega} = \frac{\mu^2}{4\pi^2\hbar^4}\left|\int e^{-ik\cdot r'}V(r')\psi(r')d^3r'\right|^2 = \frac{-\mu_2}{2\pi^2\hbar^4}\left|\langle\phi|V|\psi\rangle\right|^2$$

# Using the first Born approximation

If the potential is weak, the incident plane wave is only a little distorted and the scattered wave is also a plane wave. That's the assumption behind the first Born approximation, which you take a look at here. So if you make the assumption that the potential is weak, you can determine from the equation

$$\psi_1(r) = \phi_{inc} - \frac{\mu}{2\pi\hbar^2} \int \frac{e^{ik|r-r_1|}}{|r-r_1|} V(r_1) \psi_0(r_1) d^3 r_1 \quad \text{that}$$

$$\psi(r) \approx \phi_{inc} - \frac{\mu}{2\pi\hbar^2} \int \frac{e^{ik|r-r'|}}{|r-r'|} V(r') \phi_{inc}(r') d^3 r'$$

Okay, so what is $f(\theta, \phi)$? Well

$$f(\theta,\phi) = \frac{-\mu_2}{2\pi\hbar^2} \int e^{-ik\cdot r'} V(r') \psi(r') d^3 r'$$

And this equals the following, where $q = k_0 - k$:

$$f(\phi,\theta) = \frac{-\mu_2}{2\pi\hbar^2} \int e^{-ik\cdot r'} V(r') \psi(r') d^3 r' = \frac{-\mu}{2\pi\hbar^2} \int e^{-iq\cdot r'} V(r') d^3 r'$$

And because $\dfrac{d\sigma(\phi,\theta)}{d\Omega} = |f(\phi,\theta)|^2$ , you have

$$\frac{d\sigma(\phi,\theta)}{d\Omega} = \frac{\mu^2}{4\pi^2\hbar^4} \left| \int e^{-iq\cdot r'} V(r') d^3 r' \right|^2$$

When the scattering is elastic, the magnitude of $k$ is equal to the magnitude of $k_0$, and you have

$$q = |k_0 - k| = 2k \sin(\theta/2)$$

where $\theta$ is the angle between $k_0$ and $k$.

In addition, if you say that $V(r)$ is spherically symmetric, and you can choose the $z$ axis along $w$, then $q - r' = qr' \cos\theta'$, so

$$f(\theta,\phi) = \frac{-\mu}{2\pi\hbar^2} \int e^{-iq\cdot r'} V(r') d^3 r' = \frac{-\mu}{2\pi\hbar^2} \int_0^\infty r'^2 V(r') dr' \int_0^\pi e^{-iqr'\cos\theta'} \sin\theta' d\theta' \int_0^{2\pi} d\phi'$$

That equals

$$f(\phi,\theta) = \frac{-\mu}{2\pi\hbar^2} \int_0^\infty r'^2 V(r') dr' \int_0^\pi e^{-iqr'\cos\theta'} \sin\theta' d\theta' \int_0^{2\pi} d\phi' = \frac{-2\mu}{q\hbar^2} \int_0^\infty r' V(r') \sin(qr') dr'$$

Because $\dfrac{d\sigma(\phi,\theta)}{d\Omega} = \left|f(\phi,\theta)\right|^2$ , you know that $\dfrac{d\sigma(\phi,\theta)}{d\Omega} = \left|f(\phi,\theta)\right|^2$ , so

$$\frac{d\sigma(\phi,\theta)}{d\Omega} = \frac{-4\mu^2}{q^2\hbar^4}\left|\int_0^\infty r'V(r')\sin(qr')dr'\right|^2$$

You've come far in this chapter — from the Schrödinger equation all the way through the Born approximation, and now to the preceding equation for weak, spherically symmetric potentials. How about you put this to work with some concrete numbers?

# Putting the Born approximation to work

In this section, you find the differential cross section for two electrically charged particles of charge $Z_1 e$ and $Z_2 e$. Here, the potential looks like this:

$$V(r) = \frac{Z_1 Z_2 e^2}{r}$$

So here's what the differential cross section looks like in the first Born approximation:

$$\frac{d\sigma(\phi,\theta)}{d\Omega} = \frac{-4Z_1^{\,2}Z_2^{\,2}e^4\mu^2}{q^2\hbar^4}\left|\int_0^\infty \sin(qr')dr'\right|^2$$

And because $\int_0^\infty \sin(qr')dr' = \dfrac{1}{q}$ , you know that

$$\frac{d\sigma(\phi,\theta)}{d\Omega} = \frac{4Z_1^{\,2}Z_2^{\,2}e^4\mu^2}{q^4\hbar^4}$$

And because $q = 2k\sin(\theta/2)$, the following is true:

$$\frac{d\sigma(\phi,\theta)}{d\Omega} = \frac{4Z_1^{\,2}Z_2^{\,2}e^4\mu^2}{q^4\hbar^4} = \frac{Z_1^{\,2}Z_2^{\,2}e^4}{16E^2}\sin^{-4}\left(\frac{\theta}{2}\right)$$

where E is the kinetic energy of the incoming particle: $E = \dfrac{\hbar^2 k^2}{2\mu}$ .

Now get more specific; say that you're smashing an alpha particle, $Z_1 = 4$, against a gold nucleus, $Z_2 = 79$. If the scattering angle in the lab frame is 60°, what is it in the center-of-mass frame?

The ratio of the particles' mass in this case, $m_1/m_2$, is 0.02, so the scattering angle in the center-of-mass frame, $\theta$, is the following, where $\theta_{lab} = 60°$:

$$\tan\theta_{lab} = \frac{\sin\theta}{\cos\theta + m_1 / m_2}$$

Solving that equation for $\theta$ gives you $\theta = 61°$. So what's the cross section for this scattering angle? Take a look:

$$\frac{d\sigma(\phi,\theta)}{d\Omega} = \frac{4Z_1^2 Z_2^2 e^4 \mu^2}{q^4 \hbar^4} = \frac{Z_1^2 Z_2^2 e^4}{16E^2} \sin^{-4}\left(\frac{\theta}{2}\right)$$

Plugging in the numbers if the incident alpha particle's energy is 8 MeV gives you the following:

$$\frac{d\sigma(\phi,\theta)}{d\Omega} = 3.1 \times 10^{-29}\, m^2$$

That's the size of the target — the cross section — you have to hit to create the scattering angle seen.

# Part VI
# The Part of Tens

The 5th Wave                                                    By Rich Tennant

"What exactly are we saying here?"

# In this part . . .

I let quantum physics off the leash in this part, and it goes wild. You get to see the ten best online tutorials here, as well as ten major triumphs of quantum physics. Researchers created quantum physics because of the need to handle issues such as the wave-particle duality, the uncertainty principle, and the photoelectric effect, and you relive those triumphs here.

# Chapter 13

# Ten Quantum Physics Tutorials

. . . . . . . . . . . . . . . . . . . . . . . . . . . . . . . . . . . . . . . . . . . . . . . . . . . . . . . . . .

## In This Chapter

▶ Understanding basic concepts and equations

▶ Viewing illustrations and animations

. . . . . . . . . . . . . . . . . . . . . . . . . . . . . . . . . . . . . . . . . . . . . . . . . . . . . . . . . .

*W*hen scientists start mixing talk of dice, billiard balls, and a possibly undead cat-in-a-box, you know you're dealing with a challenging subject. Luckily, you can find plenty of online tutorials, some of them featuring animation, to help you wrap your brain around quantum physics. This chapter presents a good starter list.

## An Introduction to Quantum Mechanics

www.chemistry.ohio-state.edu/betha/qm

*What is a wave function? What is an orbital?: An Introduction to Quantum Mechanics* comes from Neal McDonald, Midori Kitagawa-DeLeon, Anna Timasheva, Heath Hanlin, Zil Lilas, and Sherwin J. Singer at The Ohio State University. This site includes tutorials on probability, particles versus waves, wave functions, and more, including Shockwave-based sound (though if you don't have Shockwave installed, that's not a problem).

## Quantum Mechanics Tutorial

www.gilestv.com/tutorials/quantum.html

This cool tutorial is one of the Flash-animated *Modern Physics Tutorials* by Giles Hogben. Extensively illustrated, this tutorial probes questions such as wave-particle duality and offers a good general introduction to quantum physics.

# Grains of Mystique: Quantum Physics for the Layman

www.faqs.org/docs/qp

This site provides good historical and experimental background info — and they've documented their sources and made some attempts at peer review.

# Quantum Physics Online Version 2.0

www.quantum-physics.polytechnique.fr/index.html

This is a cool set of programs that run in your browser, giving simulations of various quantum physics experiments. It's by Manuel Joffre, Jean-Louis Basdevant, and Jean Dalibard of the École Polytechnique in France. Look for information on wave mechanics, quantization, quantum superposition, and spin $1/2$.

# Todd K. Timberlake's Tutorial

facultyweb.berry.edu/ttimberlake/qchaos/qm.html

This tutorial is by Todd K. Timberlake, assistant professor of the Department of Physics, Astronomy, & Geology of Berry College in Georgia. It's a fairly brief but well-written introduction to the ideas of quantum mechanics.

# Physics 24/7's Tutorial

www.physics247.com/physics-tutorial/quantum-physics-billiards.shtml

This is a text-based tutorial from Physics 24/7. It includes material on quanta, the uncertainty principle, and quantum tunneling (as well as some ads).

# Stan Zochowski's PDF Tutorials

www.cmmp.ucl.ac.uk/~swz/courses/SM355/SM355.html

Stan Zochowski, from the department of Physics & Astronomy at University College London, put together these PDF-based tutorials on quantum physics. These are tutorial handouts for a Quantum Mechanics course at the University College, and they serve as an excellent introduction to quantum physics.

# Quantum Atom Tutorial

www.colorado.edu/physics/2000/quantumzone/index.html

This is a fun, cartoon-centric tutorial on the quantum nature of the atom from the University of Colorado Physics 2000 project.

# College of St. Benedict's Tutorial

www.physics.csbsju.edu/QM/Index.html

This is a comprehensive quantum physics tutorial from the College of St. Benedict. It's a good, more serious, text and equations-based tutorial with plenty of illustrations.

# A Web-Based Quantum Mechanics Course

electron6.phys.utk.edu/qm1/Modules.htm

This one's from the University of Tennessee, and it's an extensive online course in quantum physics. It includes modules on square potentials, harmonic oscillators, angular momentum, spin, and so on.

# Chapter 14

# Ten Quantum Physics Triumphs

*In This Chapter*

▶ Explaining unexpected results

▶ Identifying characteristics of the quantum world

▶ Developing new models

*Q*uantum physics has been very successful in explaining many physical phenomena, such as wave-particle duality. In fact, quantum physics was created to explain physical measurements that classical physics couldn't explain. This chapter is about ten triumphs of quantum physics, and it points you to resources on the Web that examine those triumphs for further information.

## Wave-Particle Duality

Is that particle a wave? Or is that wave a particle? That's one of the questions that quantum physics was created to solve, because particles exhibited wave-like properties in the lab, whereas waves exhibited particle-like properties.

These Web sites offer more insight:

✔ www.gilestv.com/tutorials/quantum.html

✔ www.physics247.com/physics-tutorial/quantum-physics-billiards.shtml

## The Photoelectric Effect

Another founding pillar of quantum physics was explaining the *photoelectric effect,* in which experimenters shone light on a metal. No matter how strong the light, the energy of ejected electrons from the metal didn't rise. It turns out that the energy of electrons goes up with the frequency of the light, not its intensity — which gives support to the light as a stream of discrete photons theory.

For more info on the photoelectric effect, check out `www.gilestv.com/tutorials/quantum.html`.

# Postulating Spin

The Stern-Gerlach experiment results couldn't be explained without postulating spin, another triumph of quantum physics. This experiment sent electrons through a magnetic field, and the classical prediction is that the electron stream would create one spot of electrons on a screen — but there were two (corresponding to the two spins, up and down).

This Web site has more info: `electron6.phys.utk.edu/qm1/modules/m9/spin.htm`.

# Differences between Newton's Laws and Quantum Physics

In classical physics, bound particles can have any energy or speed, but that's not true in quantum physics. And in classical physics, you can determine both the position and momentum of particles exactly, which isn't true in quantum physics (thanks to the Heisenberg uncertainty principle). And in quantum physics, you can superimpose states on each other, and have particles tunnel into areas that would be classically impossible.

You can find a nice discussion of the differences between classical and quantum physics at `facultyweb.berry.edu/ttimberlake/qchaos/qm.html`.

# Heisenberg Uncertainty Principle

One of the triumphs of quantum physics is the Heisenberg uncertainty principle: Heisenberg theorized that you can't simultaneously measure a particle's position and momentum exactly. This is one of the central theories that has destroyed classical physics.

Here's where you can find one of the best Web discussions on this topic: `www.physics247.com/physics-tutorial/quantum-physics-billiards.shtml`.

# Quantum Tunneling

How can particles go where, classically, they don't have enough energy to go? For example, how can an electron with energy E go into an electrostatic field where you need to have more than energy E to penetrate? The answer was postulated with quantum tunneling, and you can find more information about that at

`www.physics247.com/physics-tutorial/quantum-physics-billiards.shtml`.

# Discrete Spectra of Atoms

Modeling the quantized nature of atoms and orbitals is another triumph of quantum physics. It turns out that electrons can't have any old energy in an atom, but are only allowed particular *quantized* energy levels — and that was one of the foundations of quantum physics.

For a lot more on the topic, visit `www.colorado.edu/physics/2000/quantumzone/index.html`.

# Harmonic Oscillator

Quantizing harmonic oscillators on the micro level was another triumph of quantum physics. Classically, harmonic oscillators can have any energy — but not quantum mechanically. And guess which one was right?

Read all about it here:

- `www.physics.csbsju.edu/QM/Index.html`
- `electron6.phys.utk.edu/qm1/modules/module8.htm`

# Square Wells

Like harmonic oscillators, quantizing particles bound in square wells at the micro level was another triumph for quantum physics. Classically, particles in square wells can have any energy, but quantum physics says you can only have certain allowed energies.

There's plenty on the Web about it, including these two good treatments:

- ✔ www.physics.csbsju.edu/QM/Index.html
- ✔ electron6.phys.utk.edu/qm1/modules/module2.htm

# Schrödinger's Cat

Schrödinger's Cat is a thought experiment that details some problems that arise in the macro world from thinking of the spin of electrons as completely non-determined until you measure them. For example, if you know the spin of one of a pair of newly-created electrons, you know the other has to have the opposite spin. So if you separate two electrons by light years and then measure the spin of one electron, does the other electron's spin suddenly snap to the opposite value — even at a distance that would take a signal from the first electron years to cover? Tricky stuff!

For more, take a look at www.gilestv.com/tutorials/quantum.html.

# Glossary

................................................................

*H*ere's a glossary of common quantum physics terms:

**amplitude:** The maximum amount of displacement of an oscillating particle.

**angular momentum:** The product of the distance a particle is from a certain point and its momentum measured with respect to the point.

**annihilation operator:** An operator that lowers the energy level of an eigenstate by one level.

**anti-Hermitian:** The value you get when you take the Hermitian adjoint of an expression and get the same thing back with a negative sign in front of it.

**black body:** A body that absorbs all radiation and radiates it all away.

**Bohr radius:** The average radius of an electron's orbit in a hydrogen atom, about $10^{-10}$ meters.

**bound state:** A state in which a particle isn't free to travel to infinity.

**bosons:** Particles with integer spins, including photons, pi mesons, and so on.

**bra-ket notation:** Abbreviating the matrix form of a state vector as a *ket,* or $|\psi>$, and abbreviating the ket's complex conjugate, or *bra,* as $<\psi|$.

**center-of-mass frame:** In scattering theory, the frame in which the center of mass is stationary and the particles head toward each other and collide. *See also* lab frame.

**central potential:** A spherically symmetrical potential.

**commute:** Two operators commute with each other if their commutator is equal to zero. The *commutator* of operators A and B is $[A, B] = AB - BA$.

**complex conjugate:** The number you get by negating the imaginary part of a complex number. The * symbol indicates a complex conjugate.

**Compton effect:** An increase of wavelength, depending on the scattering angle, that occurs after incident light hits an electron at rest.

**conservation of energy:** The law of physics that says the energy of a closed system doesn't change unless external influences act on the system.

**creation operator:** An operator that raises the energy level of an eigenstate by one level.

**current density:** *See* incident flux.

**electron volts (eV):** The amount of energy one electron gains falling through 1 volt.

**diagonalize:** Writing a matrix so that the only nonzero elements appear along the matrix's diagonal.

**differential cross section:** In scattering theory, the cross section for scattering a particle to a specific solid angle; it's like a bull's-eye.

**Dirac's constant:** Planck's constant ($h = 6.626 \times 10^{-34}$ Joule-seconds) divided by $2\pi$. It's represented by an $h$ with a bar going through it.

**Dirac notation:** *See* bra-ket notation.

**eigenvalue:** A complex constant that represents the change in magnitude of a vector.

**eigenvector:** A vector that changes in magnitude but not direction after you apply an operator.

**elastic collision:** A collision in which kinetic energy is conserved.

**electric field:** The force on a positive test charge per Coulomb due to other electrical charges.

**electron:** A negatively charged particle with half-integer spin.

**emissivity:** A property of a substance showing how well it radiates.

**energy:** The ability of a system to do work.

**energy degeneracy:** The number of states that have the same energy.

**energy well:** *See* potential well.

**expectation value:** The most probable value an operator will return.

**fermions:** Particles with half-integer spin, including electrons, protons, neutrons, quarks, and so on.

**frequency:** The number of cycles of a periodic occurrence per second.

**Hamiltonian:** An operator for the total energy of a particle, both kinetic and potential.

**Heisenberg uncertainty principle:** *See* uncertainty principle.

**Hermitian adjoint:** A value, represented as $A^\dagger$, that replaces complex numbers with their complex conjugates, swaps bras and kets, and replaces operators with their Hermitian operators.

**Hermetian operator:** Operators that are equal to their Hermitian adjoints; in other words, an operator is Hermitian if $A^\dagger = A$.

**incident flux:** The number of incident particles per unit area per unit time.

**inelastic collision:** A collision in which kinetic energy isn't conserved.

**intensity (wave):** The time-averaged rate of energy transmitted by a wave per unit of area.

**Joule:** The MKS unit of energy — one Newton-meter.

**ket:** *See* bra-ket notation.

**kinetic energy:** The energy of an object due to its motion.

**lab frame:** In scattering theory, the frame in which one particle is incident on a particle at rest and hits it. *See also* center-of-mass frame.

**Laplacian:** An operator, represented by $\Delta$, that you use to find the Hamiltonian.

**magnetic field:** The force on a moving positive test charge, per Coulomb, from magnets or moving charges.

**magnitude:** The size or length associated with a vector (vectors are made up of a direction and a magnitude).

**mass:** The property that makes matter resist being accelerated.

**momentum:** The product of mass times velocity, a vector.

**MKS system:** The measurement system that uses meters, kilograms, and seconds.

**Newton:** The MKS unit of force — one kilogram-meter per second$^2$.

**normalized function:** A function in which the probability adds up to 1.

**orbitals:** Different angular momentum states of an electron, represented as subshells in atomic structure.

**orthogonal:** Two kets, $|\psi>$ and $|\phi>$, for which $<\psi|\phi> = 0$.

**orthonormal:** Two kets, $|\psi>$ and $|\phi>$, that meet the following conditions: $<\psi|\phi> = 0$; $<\psi|\psi> = 1$; and $<\phi|\phi> = 1$.

**oscillate:** To move or swing side to side regularly.

**pair annihilation:** The conversion of an electron and positron into pure light.

**pair production:** The conversion of a high-powered photon into an electron and positron.

**particle:** A discrete piece of matter.

**Pauli exclusion principle:** The idea that no two electrons can occupy the same state in a single atom.

**period:** The time it takes for one complete cycle of a repeating event.

**perturbation:** A stimulus mild enough that you can calculate the resulting energy levels and wave functions as corrections to the fundamental energy levels and wave functions of the unperturbed system.

**photoelectric effect:** A result in which the kinetic energy of electrons emitted from a piece of metal depends only on the frequency — not the intensity — of the incident light.

**photon:** A quantum of electromagnetic radiation. An elementary particle that is its own antiparticle.

**pi meson:** A subatomic particle that helps hold the nucleus of an atom together.

**Planck's constant:** A universal constant, $h$, that describes the relationship between the energy and frequency of a photon. It equals $6.626 \times 10^{-34}$ Joule-seconds.

**positron:** A positively charged anti-electron.

**potential barrier:** A potential step of limited extent; an electron may be able to tunnel through the barrier and come out the other side.

**potential energy:** An object's energy because of its position when a force is acting on it or its internal configuration.

**potential step:** A region in which the energy potential forms a stair shape; a particle striking the step may be reflected or transmitted.

**potential well:** A region in which there's a dip in the energy potential threshold; particles without enough energy to overcome the barrier can become trapped in the well, unable to convert the potential energy to kinetic.

**power:** The rate of change in a system's energy.

**probability amplitude:** The square root of the probability that a particle will occupy a certain state.

**probability density:** The likelihood that a particle will occupy a particular position or have a particular momentum.

**quantized:** Coming in discrete values.

**quark:** Particles that combine with antiquarks to form protons, neutrons, and so on.

**radian:** The MKS unit of angle — $2\pi$ radians are in a circle.

**radiation:** A physical mechanism that transports heat and energy as electromagnetic waves.

**scalar:** A simple number (without a direction, which a vector has).

**Schrödinger equation:** A wave function that describes how energies and probable locations of electrons change over time.

**simple harmonic motion:** Repetitive motion where the restoring force is proportional to the displacement.

**spherical coordinates:** Coordinates that indicate location using two angles and the length of a radius vector.

**spin:** The intrinsic angular momentum of an electron, classified as up or down.

**synchrotron:** A type of circular particle accelerator.

**state vector:** A vector that gives the probability amplitude that particles will be in their various possible states.

**threshold frequency:** If you shine light below this frequency on metal, no electrons are emitted.

**total cross section:** In scattering theory, the cross section for any kind of particle scattering, through any angle.

**tunneling:** The phenomenon where particles can get through regions that they're classically forbidden to go.

**ultraviolet catastrophe:** The failure of the Raleigh-Jeans Law to explain black-body radiation at high frequencies.

**uncertainty principle:** A principle that says it's impossible to know an object's exact momentum and position.

**vector:** A mathematical construct that has both a magnitude and a direction.

**velocity:** The rate of change of an object's position, expressed as a vector whose magnitude is speed.

**volt:** The MKS unit of electrostatic potential — one Joule per Coulomb.

**wave:** A traveling energy disturbance.

**wavelength:** The distance between crests or troughs of a wave.

**wave-particle duality:** The observation that light has properties of both waves and particles, depending on the experiment.

**wave packet:** A collection of wave functions such that the wave functions interfere constructively at one location and interfere destructively (go to zero) at all other locations.

**work:** Force multiplied by the distance over which that force acts.

# Index

. . . . . . . . . . . . . . . . . . . . . . . . . . . . . . . . . . . . . . . . . . . . . .

# • F •

## BUSINESS, CAREERS & PERSONAL FINANCE

Accounting For Dummies, 4th Edition*
978-0-470-24600-9

Bookkeeping Workbook For Dummies†
978-0-470-16983-4

Commodities For Dummies
978-0-470-04928-0

Doing Business in China For Dummies
978-0-470-04929-7

E-Mail Marketing For Dummies
978-0-470-19087-6

Job Interviews For Dummies, 3rd Edition*†
978-0-470-17748-8

Personal Finance Workbook For Dummies*†
978-0-470-09933-9

Real Estate License Exams For Dummies
978-0-7645-7623-2

Six Sigma For Dummies
978-0-7645-6798-8

Small Business Kit For Dummies,
2nd Edition*†
978-0-7645-5984-6

Telephone Sales For Dummies
978-0-470-16836-3

## BUSINESS PRODUCTIVITY & MICROSOFT OFFICE

Access 2007 For Dummies
978-0-470-03649-5

Excel 2007 For Dummies
978-0-470-03737-9

Office 2007 For Dummies
978-0-470-00923-9

Outlook 2007 For Dummies
978-0-470-03830-7

PowerPoint 2007 For Dummies
978-0-470-04059-1

Project 2007 For Dummies
978-0-470-03651-8

QuickBooks 2008 For Dummies
978-0-470-18470-7

Quicken 2008 For Dummies
978-0-470-17473-9

Salesforce.com For Dummies,
2nd Edition
978-0-470-04893-1

Word 2007 For Dummies
978-0-470-03658-7

## EDUCATION, HISTORY, REFERENCE & TEST PREPARATION

African American History For Dummies
978-0-7645-5469-8

Algebra For Dummies
978-0-7645-5325-7

Algebra Workbook For Dummies
978-0-7645-8467-1

Art History For Dummies
978-0-470-09910-0

ASVAB For Dummies, 2nd Edition
978-0-470-10671-6

British Military History For Dummies
978-0-470-03213-8

Calculus For Dummies
978-0-7645-2498-1

Canadian History For Dummies, 2nd Edition
978-0-470-83656-9

Geometry Workbook For Dummies
978-0-471-79940-5

The SAT I For Dummies, 6th Edition
978-0-7645-7193-0

Series 7 Exam For Dummies
978-0-470-09932-2

World History For Dummies
978-0-7645-5242-7

## FOOD, GARDEN, HOBBIES & HOME

Bridge For Dummies, 2nd Edition
978-0-471-92426-5

Coin Collecting For Dummies, 2nd Edition
978-0-470-22275-1

Cooking Basics For Dummies, 3rd Edition
978-0-7645-7206-7

Drawing For Dummies
978-0-7645-5476-6

Etiquette For Dummies, 2nd Edition
978-0-470-10672-3

Gardening Basics For Dummies*†
978-0-470-03749-2

Knitting Patterns For Dummies
978-0-470-04556-5

Living Gluten-Free For Dummies†
978-0-471-77383-2

Painting Do-It-Yourself For Dummies
978-0-470-17533-0

## HEALTH, SELF HELP, PARENTING & PETS

Anger Management For Dummies
978-0-470-03715-7

Anxiety & Depression Workbook
For Dummies
978-0-7645-9793-0

Dieting For Dummies, 2nd Edition
978-0-7645-4149-0

Dog Training For Dummies, 2nd Edition
978-0-7645-8418-3

Horseback Riding For Dummies
978-0-470-09719-9

Infertility For Dummies†
978-0-470-11518-3

Meditation For Dummies with CD-ROM,
2nd Edition
978-0-471-77774-8

Post-Traumatic Stress Disorder For Dummies
978-0-470-04922-8

Puppies For Dummies, 2nd Edition
978-0-470-03717-1

Thyroid For Dummies, 2nd Edition†
978-0-471-78755-6

Type 1 Diabetes For Dummies*†
978-0-470-17811-9

* Separate Canadian edition also available
† Separate U.K. edition also available

## INTERNET & DIGITAL MEDIA

**AdWords For Dummies**
978-0-470-15252-2

**Blogging For Dummies, 2nd Edition**
978-0-470-23017-6

**Digital Photography All-in-One
Desk Reference For Dummies, 3rd Edition**
978-0-470-03743-0

**Digital Photography For Dummies, 5th Edition**
978-0-7645-9802-9

**Digital SLR Cameras & Photography
For Dummies, 2nd Edition**
978-0-470-14927-0

**eBay Business All-in-One Desk Reference
For Dummies**
978-0-7645-8438-1

**eBay For Dummies, 5th Edition***
978-0-470-04529-9

**eBay Listings That Sell For Dummies**
978-0-471-78912-3

**Facebook For Dummies**
978-0-470-26273-3

**The Internet For Dummies, 11th Edition**
978-0-470-12174-0

**Investing Online For Dummies, 5th Edition**
978-0-7645-8456-5

**iPod & iTunes For Dummies, 5th Editio◄**
978-0-470-17474-6

**MySpace For Dummies**
978-0-470-09529-4

**Podcasting For Dummies**
978-0-471-74898-4

**Search Engine Optimization
For Dummies, 2nd Edition**
978-0-471-97998-2

**Second Life For Dummies**
978-0-470-18025-9

**Starting an eBay Business For Dummi◄
3rd Edition†**
978-0-470-14924-9

## GRAPHICS, DESIGN & WEB DEVELOPMENT

**Adobe Creative Suite 3 Design Premium
All-in-One Desk Reference For Dummies**
978-0-470-11724-8

**Adobe Web Suite CS3 All-in-One Desk
Reference For Dummies**
978-0-470-12099-6

**AutoCAD 2008 For Dummies**
978-0-470-11650-0

**Building a Web Site For Dummies,
3rd Edition**
978-0-470-14928-7

**Creating Web Pages All-in-One Desk
Reference For Dummies, 3rd Edition**
978-0-470-09629-1

**Creating Web Pages For Dummies,
8th Edition**
978-0-470-08030-6

**Dreamweaver CS3 For Dummies**
978-0-470-11490-2

**Flash CS3 For Dummies**
978-0-470-12100-9

**Google SketchUp For Dummies**
978-0-470-13744-4

**InDesign CS3 For Dummies**
978-0-470-11865-8

**Photoshop CS3 All-in-One
Desk Reference For Dummies**
978-0-470-11195-6

**Photoshop CS3 For Dummies**
978-0-470-11193-2

**Photoshop Elements 5 For Dummies**
978-0-470-09810-3

**SolidWorks For Dummies**
978-0-7645-9555-4

**Visio 2007 For Dummies**
978-0-470-08983-5

**Web Design For Dummies, 2nd Editi◄**
978-0-471-78117-2

**Web Sites Do-It-Yourself For Dummi◄**
978-0-470-16903-2

**Web Stores Do-It-Yourself For Dummi◄**
978-0-470-17443-2

## LANGUAGES, RELIGION & SPIRITUALITY

**Arabic For Dummies**
978-0-471-77270-5

**Chinese For Dummies, Audio Set**
978-0-470-12766-7

**French For Dummies**
978-0-7645-5193-2

**German For Dummies**
978-0-7645-5195-6

**Hebrew For Dummies**
978-0-7645-5489-6

**Ingles Para Dummies**
978-0-7645-5427-8

**Italian For Dummies, Audio Set**
978-0-470-09586-7

**Italian Verbs For Dummies**
978-0-471-77389-4

**Japanese For Dummies**
978-0-7645-5429-2

**Latin For Dummies**
978-0-7645-5431-5

**Portuguese For Dummies**
978-0-471-78738-9

**Russian For Dummies**
978-0-471-78001-4

**Spanish Phrases For Dummies**
978-0-7645-7204-3

**Spanish For Dummies**
978-0-7645-5194-9

**Spanish For Dummies, Audio Set**
978-0-470-09585-0

**The Bible For Dummies**
978-0-7645-5296-0

**Catholicism For Dummies**
978-0-7645-5391-2

**The Historical Jesus For Dummies**
978-0-470-16785-4

**Islam For Dummies**
978-0-7645-5503-9

**Spirituality For Dummies,
2nd Edition**
978-0-470-19142-2

## NETWORKING AND PROGRAMMING

**ASP.NET 3.5 For Dummies**
978-0-470-19592-5

**C# 2008 For Dummies**
978-0-470-19109-5

**Hacking For Dummies, 2nd Edition**
978-0-470-05235-8

**Home Networking For Dummies, 4th Edition**
978-0-470-11806-1

**Java For Dummies, 4th Edition**
978-0-470-08716-9

**Microsoft® SQL Server™ 2008 All-in-One
Desk Reference For Dummies**
978-0-470-17954-3

**Networking All-in-One Desk Reference
For Dummies, 2nd Edition**
978-0-7645-9939-2

**Networking For Dummies,
8th Edition**
978-0-470-05620-2

**SharePoint 2007 For Dummies**
978-0-470-09941-4

**Wireless Home Networking
For Dummies, 2nd Edition**
978-0-471-74940-0

## OPERATING SYSTEMS & COMPUTER BASICS

**Mac For Dummies, 5th Edition**
78-0-7645-8458-9

**Laptops For Dummies, 2nd Edition**
78-0-470-05432-1

**Linux For Dummies, 8th Edition**
78-0-470-11649-4

**MacBook For Dummies**
78-0-470-04859-7

**Mac OS X Leopard All-in-One Desk Reference For Dummies**
78-0-470-05434-5

**Mac OS X Leopard For Dummies**
978-0-470-05433-8

**Macs For Dummies, 9th Edition**
978-0-470-04849-8

**PCs For Dummies, 11th Edition**
978-0-470-13728-4

**Windows® Home Server For Dummies**
978-0-470-18592-6

**Windows Server 2008 For Dummies**
978-0-470-18043-3

**Windows Vista All-in-One Desk Reference For Dummies**
978-0-471-74941-7

**Windows Vista For Dummies**
978-0-471-75421-3

**Windows Vista Security For Dummies**
978-0-470-11805-4

## SPORTS, FITNESS & MUSIC

**Coaching Hockey For Dummies**
78-0-470-83685-9

**Coaching Soccer For Dummies**
78-0-471-77381-8

**Fitness For Dummies, 3rd Edition**
78-0-7645-7851-9

**Football For Dummies, 3rd Edition**
78-0-470-12536-6

**GarageBand For Dummies**
978-0-7645-7323-1

**Golf For Dummies, 3rd Edition**
978-0-471-76871-5

**Guitar For Dummies, 2nd Edition**
978-0-7645-9904-0

**Home Recording For Musicians For Dummies, 2nd Edition**
978-0-7645-8884-6

**iPod & iTunes For Dummies, 5th Edition**
978-0-470-17474-6

**Music Theory For Dummies**
978-0-7645-7838-0

**Stretching For Dummies**
978-0-470-06741-3

# Get smart @ dummies.com®

- **Find a full list of Dummies titles**
- **Look into loads of FREE on-site articles**
- **Sign up for FREE eTips e-mailed to you weekly**
- **See what other products carry the Dummies name**
- **Shop directly from the Dummies bookstore**
- **Enter to win new prizes every month!**

**Separate Canadian edition also available**
**Separate U.K. edition also available**